THE HYDROLOGY OF THE ⌣

There is probably more known about the hydrology of the UK than of any other country. This has led to UK policies and actions based on sound science being brought to the fore of international attention.

Water, and its movement through the hydrological cycle, is the key to keeping our planet fit for life. As the hydrology of the UK changes, the institutions responsible for policy, research and management have had to adapt to ensure that development of our water is carried out in an environmentally sustainable manner through balancing the direct needs of people with their indirect needs through a healthy environment.

This book assesses the changing hydrology of the UK. The contributors focus on key issues that affect the fundamental hydrological processes and have important implications for water resource management, flood risk and environmental quality. The book is divided into three sections:

- Section 1 examines the causes of change to the hydrology of the UK, including the impact of climate change, land use changes, geomorphological impacts and dam construction.
- Section 2 assesses the effects of these pressures on UK rivers, groundwater, lakes, ponds, reservoirs and wetlands, looking at issues including water quality, degradation, pollution, and protection.
- Section 3 examines the responses of government organisations responsible for planning and management of water, including the Environment Agencies, the British Hydrological Society and the growing urgency for a World Hydrology Initiative.

Change will continue to be a major feature of UK hydrology in the future. Whether this is driven by global warming, changing policies and subsidies in the European Union, demographic variations or changing expectations of water users, people will continue to influence the hydrological cycle and management, control and government policies will need to continue to respond to the demands for change. This book provides an understanding of the changing hydrology of the UK and the international scene today and looks to the needs for the future.

Mike Acreman is Head of Low Flows, Ecology and Wetlands at the Institute of Hydrology, Wallingford, Oxfordshire, UK.

THE HYDROLOGY OF THE UK

A study of change

Edited by Mike Acreman

in association with the British Hydrological Society

London and New York

First published 2000
by Routledge
11 New Fetter Lane, London EC4P 4EE

Simultaneously published in the USA and Canada
by Routledge
29 West 35th Street, New York, NY 10001

Routledge is an imprint of the Taylor & Francis Group

© 2000 Selection and editorial matter, Mike Acreman; individual chapters,
the respective contributors

Typeset in Garamond by Graphicraft Limited, Hong Kong
Printed and bound in Great Britain by
Biddles Ltd, Guildford and King's Lynn

British Library Cataloguing in Publication Data
A catalogue record for this book is available from the British Library

Library of Congress Cataloging in Publication Data
A catalog record for this book has been requested

ISBN 0–415–18760–5 (hbk)
ISBN 0–415–18761–3 (pbk)

CONTENTS

CONTENTS

CONTENTS

CONTENTS

PLATES

PLATES

FIGURES

TABLES

BOXES

NOTES ON CONTRIBUTORS

Mike Acreman is Head of Low Flows, Ecology and Wetlands at the Institute of Hydrology, and Freshwater Management Advisor to IUCN – The World Conservation Union. He is a specialist in wetland hydrology and sustainable management.

Brian Adams is Principal Hydrogeologist for the British Geological Survey. He specialises in groundwater resource assessment and issues of groundwater protection, and is a member of the International Association of Hydrogeologists Commission for Groundwater Protection.

Nigel Arnell is Reader in the Department of Geography, University of Southampton. He is a specialist in hydrological impacts of climate change, and is lead author for the chapter on water in the Third Assessment Report of the Inter-Government Panel on Climate Change.

Tony Bailey-Watts in an ecologist at the Institute of Freshwater Ecology. His main interests lie in environmental determinants, including fluxes of nutrients from land to water, of lake phytoplankton species composition and dynamics. He has headed much of the thirty years' research on the eutrophic Loch Leven (Scotland), and has had scientific involvement in the Great African Lakes.

Rick Battarbee directs the Environmental Change Research Centre at University College London. He is mainly interested in the use of recent lake sediments for hind-casting trends in lake status, especially with reference to the impacts of eutrophication, acidification and climate change.

Jeremy Biggs is Lecturer in the School of Biological and Molecular Sciences at Oxford Brookes University. He is Co-founder and Manager of Pond Action (1988–99). His special interests are the ecology and conservation of small waterbodies.

Andrew Black is Lecturer in Physical Geography at the University of Dundee, and was formerly Secretary of the Scottish Hydrological Group

(1994–97). His research interests are in flood generation, estimation and impact.

John Boardman is Co-ordinator of the MSc programme in 'Environmental Change and Management' at the Environmental Change Unit, University of Oxford, and Chairman of COST Action 623 'Soil Erosion and Global Change'. He is a specialist in soil erosion and landscape change.

Geoff Brighty is Environmental Toxicology Manager at the Environment Agency's National Centre for Ecotoxicology and Hazardous Substances, and Manager of the Agency's research programme into environmental endocrine disruptors and their impacts in the environment.

Tim Burt is Professor of Geography and Master of Hatfield College at Durham University. He is also Chairman of the Field Studies Council and former Chairman of the British Geomorphological Research Group. He is co-editor of *Hydrological Forecasting* (Wiley, 1985), *Process Studies in Hillslope Hydrology* (Wiley, 1990) and *Nitrate* (Wiley, 1993).

John Chilton is Principal Hydrogeologist for the British Geological Survey, and Hydrogeological Advisor to DFID. His specialty is groundwater quality assessment, including studies of groundwater pollution from agricultural activities.

Craig Elliott is Research Manager for the Water Group, CIRIA, and was formerly Head of the Hydro-ecological Modelling Group at the Institute of Hydrology (1994–98). He has published widely on physical habitat modelling in rivers and setting ecologically acceptable flows.

Rob Evans is Research Fellow in the Geography Department at Anglia Polytechnic University, and Research Associate in the Department of Geography, Cambridge University, specialising in soils and remote sensing. He is also an advisor on land use and the environment to the National Farmers' Union and is the author of *Soil Erosion and Its Impacts in England and Wales* (Friends of the Earth, 1996).

Karen Fisher is Senior Engineer in the Water Management Department, HR Wallingford. She is a civil engineer specialising in performance of environmental channels, especially vegetation, and maintenance.

Ian Gale is Principal Hydrogeologist for the British Geological Survey, specialising in groundwater resource assessment, artificial recharge of aquifers and low enthalpy geothermal resources.

John Gardiner MBE is Principal, Pacific North-west Branch, Philip Williams & Associates Ltd. He is also Visiting Professor at the University of Hertfordshire, and Emeritus Professor at the Flood Hazard Research Centre,

Middlesex University. He is the author/editor of many publications, including *River Projects and Conservation* (Wiley, 1991).

Angela Gurnell is Professor of Physical Geography at the University of Birmingham. She is a member of the editorial board of *Hydrological Processes*, and the editor of several books, including *Changing River Channels* (Wiley, 1995).

Ron Harriman is head of the Environmental Studies group at the Freshwater Fisheries Laboratory. His general interest is the Scottish freshwater environment as a habitat for freshwater fish. Current studies include recovery from freshwater acidification, Critical Loads applications at the catchment scale and the effects of land-use change.

Kate Heppell is a Research Fellow at University College London, specialising in the transport and fate of pesticides in the environment.

Paul José is the UK Wetlands Advisor to the Royal Society for the Protection of Birds. He is co-author of *UK Floodplains* (Westbury, 1998), *The Wet Grassland Guide* (RSPB, 1997) and *Reedbed Management* (RSPB, 1996).

Frank M. Law is Deputy Director of the Institute of Hydrology, and was President of the British Hydrological Society from 1997 to 1999. He worked until 1988 with the Binnie consulting group in the UK as Chief Hydrologist and in Australia as a Director. He is the joint author of *Water Supply*, fourth edition (Edward Arnold, 1994).

Graham Leeks is Head of Thematic Science at the Institute of Hydrology, Science Co-ordinator of the NERC Land-Ocean Interaction Study (LOIS), and Programme Manager of the Terrestrial Initiative in Global Environmental Research (TIGER). He has authored a wide range of research papers on the fluvial sediment dynamics of upland and lowland rivers.

David Lerner is Professor of Environmental Engineering at the University of Sheffield, and Leader of the Groundwater Protection and Restoration Research Group. He is also Chair of a UNESCO Working Group which is preparing a book on urban groundwater pollution.

Alex Lyle is at the Institute of Freshwater Ecology. He has some twenty years' experience in freshwater projects including physical limnology, water resources and the ecology and conservation of fish.

Terry Marsh is Leader of the National Hydrological Monitoring Programme at the Institute of Hydrology. He is the editor of the 'Hydrological Data UK' series of publications, and the author of numerous papers, reports and articles concerning hydrology, hydrometry and water resources in the UK and abroad.

Malcolm Newson is Professor of Physical Geography at the Centre for Land Use and Water Resources Research, University of Newcastle upon Tyne. He is the author of *Land, Water and Development* (Routledge, 1997) and other texts.

Enda O'Connell is Professor of Water Resources in the Department of Civil Engineering, University of Newcastle upon Tyne, specialising in numerical modelling, and also Associate Editor of *Hydrology and Earth System Science*.

John Packman is the Urban Hydrology Research Leader at the Institute of Hydrology.

Geoff Petts is Head of the School of Geography and Environmental Sciences, and Director of the Centre for Environmental Research and Training, both at the University of Birmingham. He is also a member of the Steering Committee of the UNESCO–IHP Ecohydrology Programme, and of the ICSU Scientific Committee on Water Research. He is Editor-in-Chief of *Regulated Rivers*.

Nick Reynard is Head of the Climate Change Impacts Section and lead scientist on climate studies at the Institute of Hydrology. He was formerly Co-ordinator of the Programme on Climate Change Impacts for the Department of the Environment, Transport and the Regions.

Mark Robinson is Head of Landscape Interactions at the Institute of Hydrology, and an advisor to MAFF on the impacts of drainage on river regimes. He is co-author of *Principles of Hydrology* (McGraw-Hill, 1999).

Bob Sargent is Head of Environmental Services for the Scottish Environment Protection Agency. He was formerly Chief Hydrologist for the Forth River Purification Board.

David Sear is Senior Lecturer in Physical Geography at the University of Southampton, and is an advisor to the Environment Agency on fluvial geomorphology. He has published widely on sediment transport and applied river geomorphology.

Simon Slater is a Strategic Planner at the Environment Agency. He has worked as a reviewer of the National Rivers Authority's complete Catchment Management Plans programme and is a specialist on sustainable development.

Susan Walker was Regional Water Manager for the Environment Agency until 1998, and President of the British Hydrological Society from 1993 to 1995. She is now Professor of Physical Geography at Aberdeen University.

Jim Wallace is Director of the Institute of Hydrology, and Visiting Professor in Hydrology at the University of Reading. He is also Hydrological Advisor to the UK Government and UK Permanent Representative to WMO.

John Waterworth was Head of Water Engineering for the DoE Northern Ireland Environment and Heritage Service from 1991 to 1997. He is also a former member of the International Hydrological Programme Committee of UNESCO.

Paul Whitehead is Professor of Geography at the University of Reading. He is a specialist in water quality modelling.

David Wilcock is Professor in the School of Environmental Studies, University of Ulster. He is a member of the Northern Ireland Council for Nature Conservation and the Countryside and Chairman of its Scientific Committee. His research interests are in hydrology and fluvial geomorphology.

Richard Williams is Head of the Hydrochemical Processes Section at the Institute of Hydrology. He has published widely for more than ten years on the transport of pesticides to surface and groundwaters.

Paul Younger is Reader in Water Resources at the University of Newcastle. He is a hydrogeologist and environmental engineer specialising in the characterisation and remediation of pollution problems associated with abandoned coal and metal mines.

ACKNOWLEDGEMENTS

The original concept for this book was initiated by Mike Acreman and developed with constructive comments from the National Main Committee of the British Hydrological Society, especially Angela Gurnell and Richard Davis. An underlying principle was to encompass a wide range of hydrological expertise within the UK, and the list of contributors bears witness to the attempt to achieve this. Thanks go to Sarah Lloyd and Sarah Carty at Routledge who provided great encouragement and promoted the idea for the book. A special mention goes to Heather Turner at the CEH Institute of Hydrology who converted a multitude of text formats into that which met the publisher's specifications.

All authors have donated proceeds from this book to WaterAid.

e-mail information@wateraid.org.uk
website www.wateraid.org.uk

FOREWORD

Every professional group needs to review its practice at intervals – whether to recognise achievements or to reflect on unknowns. Hydrologists take stock at least as often as others because ours is an interdisciplinary science. Indeed, hydrology ranges from the purest science to the most applied technology. It draws together earth and life sciences, engineering and social science. Some of its practitioners address immediate political concerns, while others tackle matters of theoretical consequence.

The hydrological cycle follows water in any of its forms from cloud to land to river to sea to cloud. However, hydrology could not emerge as a subject until hydrometric measurement took root after the Renaissance and especially during the Industrial Revolution. Once river flows were quantified through seasons and years, the subtleties of their different regimes challenged the comprehension of river and water supply engineers. The link to climate was pursued vigorously and subsequent stimulus was added by the modifications caused by geology, soils and vegetation. Changes superimposed by humankind through agricultural intensification and urbanisation were to follow as topics of concern.

A major change, dated to the 1970s, occurred when hydrologists took the transformation of water quality along its natural pathways as a fruitful line of enquiry. Environmental controls on that quality were seen in the acid rain phenomenon, and in diffuse pollution from land clearance and agroforestry practices. With the change came a renewed interest in the flux of sediments and chemicals from source to sea.

As the millennium comes to a close, it is noticeable that hydrologists have been major users of remote sensing, and of information technology, including Global Information Systems (GIS). Consequently, they have become not just expert in regional environmental databases but can contribute to the many global circulation models from which climate change predictions emerge. Their ability to manipulate data warehouses of spatial and time series information coincides with the advent of the CD-ROM and the World Wide Web as accessible repositories of past and present knowledge.

Water is essential to so much that is either enjoyable or utilitarian in life that it can be taken for granted until the disaster of flood or drought comes close to home. Experienced and up-to-date hydrologists are in their element in sizing potential departures from the norm. Whether they convey that knowledge sufficiently well to community leaders and the public is uncertain. This book is just one part of the contribution of the British Hydrological Society to that requirement. Hopefully, it will also serve to assist a new generation of geography, environmental science and civil engineering graduates to see the attractions of a career in hydrology.

Frank Law
President BHS (1997–99)
Institute of Hydrology
Wallingford

THE CHANGING HYDROLOGY OF THE UK

Mike Acreman

Water is the lifeblood of our planet. It is fundamental to the biochemistry of all living organisms. It has been exploited by man to irrigate crops, generate hydro-electric power, drive industrial processes and to create recreational areas. The planet's ecosystems are linked and maintained by water; water aids plant growth and provides a permanent or temporary habitat for many species. Water is also a universal solvent and provides the major pathway for the flow of sediment, nutrients, pollutants and some pathogens. Through erosion, transportation and deposition by rivers, glaciers and ice-sheets, water shapes the landscape; through evaporation and condensation, it drives the energy exchange between land and atmosphere, thus controlling the Earth's climate.

The nature of the hydrological cycle means that water in the biosphere is in constant motion as precipitation, percolation, groundwater flow, runoff and evaporation. It moves within and between the various stores in clouds, soils, aquifers, rivers, lakes, wetlands and the sea. The volumes of water within these stores, and rates of exchange between them, are controlled by a multitude of factors. These include climate, land use, catchment development (such as dams and groundwater abstraction) and channel engineering (for example dredging). These factors are themselves continually changing, but over different timescales (Table I.1).

Although there is still debate about whether recent droughts and floods are the result of global warming or just natural climatic variability, past ice ages are undoubted evidence of long-term climate change. Deforestation to provide agricultural land over the past 5000 years has caused significant changes in the hydrology of the UK (in terms of both water quantity and quality) and recent land use conversion due, for example, to altered farming subsidies, ensures that such changes will continue.

The potential energy of water was recognised at an early time and the first machines of the Industrial Revolution of the eighteenth century were powered by water-wheels, which required the construction of thousands of

Table I.1 Hydrological changes over different timescales.

Timescale	Types of change	Example of impact
Millions of years	Continental drift	Altered rainfall and runoff
Thousands of years	Change of earth's orbit	Ice ages
	Tectonic uplift	Increased erosion
Hundreds of years	Man-induced greenhouse effects	Increase in flood risk
	Deforestation	Decline in total runoff
	Industrial decline	Rising groundwater, cleaner rivers
Tens of years	Groundwater abstraction	Falling groundwater levels
	Pesticide application	Groundwater pollution
One year	Climatic fluctuations	Droughts
	Change in crop type	Increased evaporation
One day	Dam operation	Fewer downstream spates
	Chemical spills	Fish kills

small dams and mill-lakes. Rapid population growth rates and expectations for higher standards of living during the twentieth century have led to increased demand for reliable public water supply and sanitation, food security, reduced flood risk, better industrial goods and cheap electricity. Meeting these demands has led to conversion of wetlands to arable land, increased river management, groundwater abstraction, the expanded use of agricultural chemicals and increased discharge of effluent. As a result, the hydrology of the UK has been altered significantly. The Environment Agency's River Habitat Survey (Raven *et al.*, 1997) showed that less than 10% of sample river reaches were free from structural modification of channel and banks. Gustard *et al.* (1992) found that 442 catchments out of 1366 studied had flow regimes that differed by more that 50% from their natural condition (720 differed by more than 20%). English Nature (1997) suggests that since 1930, 64% of the wet grassland in Thames valley, 48% in Romney Marsh and 37% in Norfolk/Suffolk Broads has been lost. Many hydrological changes have been positive with improved public water supply and locally reduced flood risk. By contrast, others have been negative and have caused degradation of our aquatic ecosystems.

As the hydrology of the UK changes, the institutions responsible for policy, research and management have had to adapt to ensure that development of our water is carried out in an environmentally sustainable manner through balancing the direct, often short-term, needs of people with their indirect, often long-term, needs through a healthy environment. Local Environment Action Plans fostered by the Environment Agency (in England and Wales), Assess Management Plans of water companies and Agenda 21 reports (from County Councils/NGOs and consortia) are examples of this new thinking. One problem is that because of natural changes there is no abso-

Table I.2 The changing hydrology of the UK structured in terms of the DPSIR (driving forces, pressures, state, impacts, responses) model.

Driving forces	Pressures (causes of hydrological change)	Effects (past, current and future state of UK hydrology)	Impacts (positive and negative results of change)	Responses (organisations and strategies for planning and management)
• Greenhouse effect • Rising population • EU subsidies and directives	• Climate change • Land use change • Channel modification • Catchment scale hydrological changes	• Rivers (quantity) • Rivers (quality) • Groundwater • Lakes and ponds • Wetlands	• Degraded ecosystems • Loss of water rights • Increased reliability of water supply	• Responsibilities and strategies of UK organisations • Planning and managing for the future • Role of the British Hydrological Society • Future UK hydrological research

lute baseline against which to measure anthropogenic impacts. Furthermore, many aspects of a managed rivers, lakes and wetlands are now accepted as normal and preferred by society to truly natural hydrological features. Consequently, management and restoration objectives are partly governed by social choice.

This book assesses the changing hydrology of the United Kingdom of Great Britain (England, Scotland and Wales) and Northern Ireland (at a time when the federal nature of government administration is being strengthened). It does not attempt to cover every issue of change, the reasons for it and the responses to it, but rather focuses on key issues that affect the fundamental hydrological processes and have important implications for water resource management, flood risk and environmental quality.

The book is divided into three sections. Section 1 describes the causes of change to the hydrology of the UK. These are the pressures that are responsible for major alterations in the hydrological cycle (Table I.2). The original driving forces from outside the hydrological cycle, such as carbon emissions, demographic changes, and European Union (EU) farming subsidies are beyond the scope of the book, as are the ultimate impacts on ecosystems and water supply issues.

A major cause of hydrological change is climate, so Chapter 1 explores the potential effects of climate change on hydrological characteristics in the UK over the coming decades, considers uncertainty in predictions and examines implications for water management. Chapter 2 describes land use changes, including agriculture, forestry, urban development, and their implications for hydrology, soil erosion and sediment transport. Chapter 3 considers instream changes such as direct channel modification and associated hydrological and geomorphological impacts. Chapter 4 focuses on catchment-scale hydrological changes including abstraction, inter-basin transfers and dam construction.

Section 2 assesses the effects of these pressures on UK rivers (in terms of both quantity and quality), groundwater, lakes, ponds and reservoirs and wetlands. Chapter 5 considers the spatial and temporal variations in flow patterns across the UK in terms of low flows, flooding risk and the implications for aquatic ecology. Chapter 6 explores the major issues of water quality in our rivers, focusing on nutrients (nitrates and phosphates), pesticides and endocrine disrupters. Chapter 7 discusses key groundwater issues, including artificial storage and recovery – where surplus surface water is recharged artificially into an aquifer during winter for subsequent re-abstraction during the summer – pollution by nitrates and pesticides, acidic and metalliferous outflows from abandoned mine workings and rising groundwater in urban areas. Chapter 8 looks into the issues concerning the lakes and ponds of the UK highlighting the degradation and loss of ponds by intensive land use and the problems of acidification and eutrophication of lakes. The final chapter in Section 2, Chapter 9, reviews the state of the UK's wetlands, their hydrological importance and attempts to protect and restore wetlands.

Section 3 of the book deals with the responses of the government organisations responsible for planning and management of water in the UK. Chapter 10 considers the responsibilities and the strategies of UK organisations: the Environment Agency (EA) in England and Wales, the Scottish Environment Protection Agency (SEPA) and the Department of the Environment (DoE), Northern Ireland. It also outlines the role of other stakeholders in water management including the water companies, power generators, local authorities and research organisations. Chapter 11 reviews the basis for integrated catchment management and how the UK can achieve sustainability with respect to the hydrological cycle – as required in the EU Water Framework Directive. Chapter 12 focuses on the role of the British Hydrological Society (BHS) in providing a forum for debate of scientific and applied aspects of UK hydrology. It includes the Exeter Statement that has reinforced the growing urgency for a World Hydrology Initiative, an idea now before UNESCO, WMO and other key global organisations. Finally, Chapter 13 analyses the gaps in our understanding of UK hydrology and needs for future research.

There is probably more known about the hydrology of the UK than of any other country, which has led to policies and actions based on sound science. Yet, uncertainty still pervades all aspects of hydrology. Field measurements are only estimates of hydrological stores and processes and models represent further approximations of a complex world. One certainty is that change will continue to be a major feature of UK hydrology in the future. Whether this is driven ultimately by global warming, changing policies and subsidies in the European Union, demographic variations or changing expectations of water users, people will continue to influence the hydrological cycle. In some cases, this will involve more control, such as inter-basin transfers to improve water supply reliability; in other cases, it will mean restoring natural processes, such as reconnecting floodplains and rivers to reduce flood risk and enhance wildlife habitat. The present (1999) government's consultation about the need to make water licences time limited is a critical initiative. The hydrology of the UK is by and large a managed system and will increasingly be so.

References

English Nature (1997) *Wildlife and Fresh Water – An Agenda for Sustainable Management*. English Nature, Peterborough.

Gustard, A., Bullock, A. and Dixon, J. (1992) Low flow estimation in the United Kingdom. *Institute of Hydrology Report No. 108*. Institute of Hydrology, Wallingford.

Raven, P.J., Fox, P., Everard, M., Holmes, N.T.H. and Dawson, F.H. (1997) River Habitat Survey: a new system to classify rivers according to their physical character, in Boon, P.J. and Howell, D.L. (eds), *Freshwater Quality: Defining the Indefinable*. The Stationery Office, Edinburgh. pp. 215–34.

Section 1

CAUSES

1

CLIMATE CHANGE AND UK HYDROLOGY

Nigel Arnell and Nick Reynard

This chapter explores the potential effects of climate change on the hydrology of the UK over the coming decades. It reviews recent climatic variability across the UK, considers how the UK climate might change in the future and explores the implications of these possible climate changes – and the uncertainty in estimates of the future – for water management.

1.1 The context: climatic variability, climate change and global warming

A series of prolonged droughts across many parts of the UK during the 1980s and 1990s, coupled with unprecedented floods, have drawn attention to the variability in the UK weather, and more particularly to the potential effects of climate change. Since 1990, hardly a month has passed without the media asking whether the 'unusual' weather is a result of climate change, and managers within the water industry are increasingly taking the threat of climate change seriously; it is even beginning to affect investment decisions.

Weather describes the day-to-day state of the atmosphere, and climate is average weather. Climate, however, is not constant, and varies over time scales ranging from several million years, through hundreds of thousands of years (glacial/interglacial cycles), thousands of years, to decades. Although it is difficult to draw a clear distinction, climate change can be interpreted as resulting from a change in the variables forcing climate (radiation inputs, for example), whilst climatic variability reflects the inherent fluctuations in climate with no change in forcing variables. This variability may reflect the effects of periodic fluctuations in ocean currents for example (as in the case of the El Niño), or the effects of some critical threshold being crossed (as around 10 500 years ago, when the sudden influx of meltwater into the

3

north Atlantic triggered a rapid cooling, lasting for several hundred years, interrupting the post-glacial warming). Recently, however, the term 'climate change' has come to be synonymous with 'global warming'.

Global warming is the apparent increase in global average temperatures experienced over the last hundred years, and particularly since the 1960s. Since the middle of the nineteenth century, global average temperature has risen by over 0.5 °C (IPCC, 1996) and nine of the ten warmest years since 1860 have occurred between 1980 and 1997. This increase has been attributed to the rising concentration in the atmosphere of so-called greenhouse gases, such as water vapour, carbon dioxide, and methane, as a result of human activities. These gases trap radiation in the lower atmosphere, and lead to increased temperatures at the Earth's surface. Variations in solar output and the effects of volcanic eruptions (which reduce incoming radiation) also affect the radiation balance, but these influences are not large enough to account for the observed rise in global average temperature (Hadley Centre, 1997).

During the late 1980s, the apparent rise in global temperatures, coupled with the results of early simulation experiments which showed that rising concentrations of greenhouse gases could significantly affect climate, led the United Nations to establish the Intergovernmental Panel on Climate Change (IPCC). The IPCC is a collection of working groups of experts in climate science and impact analysis. It produced its Second Assessment Report during 1996, in which it was concluded that there was a discernible human influence on climate and that continued emissions of greenhouse gases would lead to some significant adverse impacts (IPCC, 1996).

However, although the IPCC concluded that global warming probably was a result of human activity, it is not possible to attribute individual events in specific locations directly to climate change. The UK has experienced droughts and floods in previous decades – the drought of the mid-1930s being particularly notable – and would continue to do so without climate change. As will be shown below, feasible future climates for the UK generally imply an increased frequency in floods and droughts (at least in some parts of the country), but that does not mean that recent events are caused by, or conclusive proof of, global warming.

1.2 Observed climatic variability across the UK over the last few decades

1.2.1 Introduction: variability in temperature and precipitation across the UK

The UK's weather is well known for its variability at all time scales from day to day, month to month and year to year. Climate – weather averaged over several years – also varies. The UK is fortunate in having long records of both rainfall and temperature at many sites across the country, and much

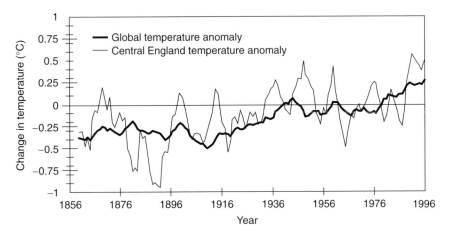

Figure 1.1 Global and UK temperature trends: 1860–1997.

effort has been expended in creating homogeneous regional and national data series (Hulme and Barrow, 1997).

The Central England Temperature (Manley, 1974; Parker *et al.*, 1992) record provides a time series of monthly temperatures back to 1659. Figure 1.1 shows the annual average Central England Temperature from 1860–1997, plotted against the global average temperature. Both are shown as five-year moving averages of the anomalies from the 1961–90 average. Even with this smoothing, the large degree of variability is evident. Despite this variability, the 1990s appear as a sustained warm period, with four of the five warmest years on the series being recorded since 1989, and 1997 ranking third in the entire 300-year series (the warmest year is 1990, with 1949 second).

Similar trends in either national or global rainfall data sets are harder to find. However, the recent past has seen a remarkable shift in the spatial and temporal rainfall patterns in the UK. There has been a tendency towards wetter winters and drier summers, with rainfall totals increasing particularly in the north and west of the UK (Marsh, 1996). This is discussed in more detail in Chapter 7. Other parts of the world have experienced changes in rainfall patterns over the last few decades, but the global average precipitation – unlike global average temperature – shows no clear pattern.

1.2.2 *Why has climate varied?*

Why have temperature and precipitation varied across the UK in the last few decades, and why have there been geographical differences in this variability? Global warming is not the entire answer (and may not actually be much of it at all yet), as climate has varied considerably over the entire instrumental record.

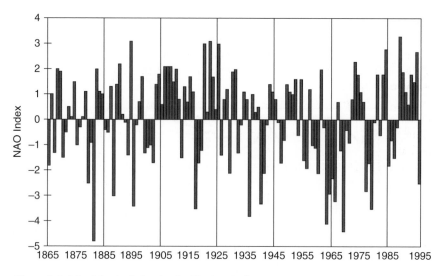

Figure 1.2 The North Atlantic Oscillation Index.

The UK's weather and climate are largely driven from the Atlantic, and are characterised by the passage of depressions from the west bringing variable amounts of rain. Occasionally, the weather is dominated by prolonged periods of high pressure centred over the continent or the mid-Atlantic. These conditions tend to produce notable weather events, such as droughts (for example, 1976) and cold, dry winters (such as 1962/63). In general, however, the weather and climate across the UK in any year are determined by the location, frequency and vigour of the depressions passing across the UK from the Atlantic. Depression tracks and properties depend on what is happening in the north Atlantic, and particularly on pressure gradients and sea surface temperatures. These vary quasi-periodically.

As long ago as the mid-eighteenth century, it was noticed that when Greenland was colder than average in winter, temperatures were above average in northern Europe, and vice versa. By the 1920s, this phenomenon was termed the North Atlantic Oscillation (NAO). It was subsequently found that the difference from year to year could be interpreted in terms of westerly airflows, and hence pressure gradients, with high winter temperatures in Europe associated with strong westerlies. The NAO can therefore be characterised by the meridional pressure gradients across the north Atlantic, and the North Atlantic Oscillation Index (NAOI) is a measure of the pressure difference between Iceland (where pressure is generally low) and the Azores (where pressure is high). Figure 1.2 shows the NAOI from 1865–1995 (Hurrell, 1995). The index has been particularly high during the 1980s and 1990s. When the NAOI is high, westerlies are more powerful and tend to bring more rain and higher temperatures to northern Europe and northern

UK (Hurrell, 1995). When the NAOI is low, depressions are weaker and continental influences are more strongly felt in the UK. Shorthouse and Arnell (1997) found strong correlations between the NAOI and streamflow across Europe, particularly in winter, with increased runoff during years with high NAOI.

The El Niño/Southern Oscillation (ENSO) is a well-studied feature of climatic variability in the southern Pacific, which has documented connections with climatic anomalies across large parts of the southern hemisphere and around the world. The effects of ENSO on climatic extremes around the Pacific are well-documented (floods along the west coast of the United States during warm ENSO years, for example), and weak ENSO signals have been found in European climate (Fraedrich and Muller, 1992; Fraedrich, 1994; Wilby, 1993). In the winter following a warm ENSO event, for example, precipitation is higher than average across northern Europe (Wilby, 1993), due to changes in the tracks of depressions, but it is difficult to separate the effects of ENSO and the NAO.

These patterns in global and regional climate are increasingly well understood, and have documented effects on multi-decadal climatic variability. Future climate change must be seen in the context of this inherent variability. Is the strengthening NAOI since the 1980s a consequence of global warming, or is it a feature of natural climatic variability controlled by long-term rhythms in the North Atlantic? How would global warming affect the North Atlantic Oscillation? These questions are currently unanswered, but are the subject of much research.

Climatic fluctuations have manifest themselves in patterns in the variability over time of river flows and groundwater recharge in the UK. These patterns are explored in Chapter 7.

1.3 Future climate

1.3.1 The enhanced greenhouse effect and global warming

The 'greenhouse effect' is not a twentieth-century phenomenon and is by no means purely a result of human activities. Without the presence of certain greenhouse gases, principally water vapour, in the atmosphere the average temperature of the Earth would be about $-15°C$, some $30°C$ cooler than it is at present.

The Earth's atmosphere is transparent to incoming solar radiation. Some of this radiation is reflected back directly by the surface of the Earth and the cloud tops, but most is absorbed by the surface, which warms as a result. This warmed surface re-emits infra-red radiation into the lower atmosphere, some of which passes straight through into space, but some is absorbed and re-emitted by so-called greenhouse gases (GHGs). The most important of these gases, by virtue of the absolute volume in the atmosphere, is water

Plate 1.1 A major source of greenhouse gas emission is from the combustion of fossil fuels (Celia Kirby).

vapour, but others include carbon dioxide (CO_2), nitrous oxides (NO_x) and methane (CH_4). This natural greenhouse effect is being enhanced due to human activities (Plate 1.1), increasing the concentrations of GHGs such as CO_2, NO_x and CH_4, and adding new and efficient GHGs such the family of chlorofluorocarbons (CFCs). Increasing concentrations of greenhouse gases such as CO_2, NO_x and CH_4 lead to increased concentrations of water vapour, through the positive feedback effect of increased temperatures, leading to increases in net radiation at the Earth's surface and temperature in the lower atmosphere.

Set against these trends, however, are the effects of sulphate aerosols. These are a result of the burning of fossil fuels, and serve to reduce temperatures. The sulphate aerosols reflect incoming solar radiation back to space, and also provide condensation nuclei for water vapour; the resulting clouds also reflect incoming solar radiation. The residence time of sulphate aerosols in the lower atmosphere can be measured in days or weeks, but they have very important effects on climate. The 'pause' in global warming during the 1940s and 1950s has been attributed to the rapid increase in sulphate aerosol emissions (IPCC, 1996). However, sulphate aerosols are also associated with air pollution, and action to improve air quality will lead to reductions in the emission of sulphate aerosols.

1.3.2   Estimating future climates

The basic physics behind the greenhouse effect and global warming is very simple, but exactly how these will affect the world climate in the future is far from certain. There are various methods for determining what these possible climate changes might be due to global warming, the most popular being the use of General Circulation Models (GCMs). GCMs seek to represent mathematically the current climate and ocean systems and hence allow experiments to be done to investigate the response of the global climate to changes in the atmospheric concentrations of greenhouse gases. These models generally run on a grid network. The UK Hadley Centre GCM, for example, has a spatial resolution of 3.75° longitude by 2.5° latitude, which produces a total of 7008 cells covering the globe.

GCMs can be used to simulate the evolution of climate in response to a gradual increase in greenhouse gas concentrations. A typical 'transient' experiment runs from pre-industrial times through to 2100, with the historical emission rates of the greenhouse gases used up to the present day. Different GCM experiments then use different assumed rates of increase of greenhouse gas concentrations for the future, and compute projections of future climate. These simulated future climates are then compared with simulated present climate to assess the potential change in climate.

There are many uncertainties with each type of approach, however. These are associated with the complexity of the global climate system and ability of the models to simulate it accurately at both global and regional scales. Current GCMs simulate well the broad features of global climate, but regional (sub-continental scale) climates may be very badly reproduced. Also, climate models tend to 'drift' gradually away from a steady state even without a change in greenhouse gas concentrations. This drift has conventionally been removed through the use of 'flux correction factors'. Other uncertainties in GCMs relate to the treatment of clouds, sea ice and the interaction between the land surface and the atmosphere.

1.3.3   Possible future climates in the UK

Climate models do not produce forecasts of future climate, but instead produce projections of potential future climate; these projections can be seen as scenarios for possible futures. Different climate models produce different projections, as do different assumptions about the rate of increase in greenhouse gas concentrations. This section describes briefly three possible climates for the UK in the 2050s, each produced in a slightly different way. They serve also as examples of how to estimate possible futures. Each is based on output from Hadley Centre GCMs. The scenarios are summarised in Table 1.1.

The first scenario is based on the HadCM1 experiment, and is the same as used by the Climate Change Impacts Review Group in its 1996 assessment

Table 1.1 Climate change scenarios.

Scenario	Experiment type	Scenario construction	Source reference
HadCM1	Transient model	Last decade of model run rescaled to 1.56°C increase from 1961–90	CCIRG (1996); Murphy and Mitchell (1994)
GG1	HadCM2 transient model	Last 30 years of model run rescaled to 1.56°C increase from 1961–90	Johns *et al.* (1997)
GGx	HadCM2 transient model: average of four runs	30 years around 2050	Johns *et al.* (1997)

of the potential impacts of climate change in the UK (CCIRG, 1996). The HadCM1 experiment assumed a compound 1% increase in greenhouse gas concentrations. The scenario for the 2050s was constructed (Hulme, 1996) in three stages. First, the spatial pattern of change in temperature, rainfall and other climatic variables was defined from the last decade of the HadCM1 simulation run, and standardised to represent change per degree Celsius of global warming. Second, the global warming expected by the 2050s under a given emission scenario was estimated using a simple one-dimensional climate model (Wigley and Raper, 1992). Finally, the spatial patterns of change in climate were rescaled by the global temperature change, to estimate the absolute magnitude of change in climate across the UK by the 2050s. This three-stage approach was adopted for two primary reasons. Simulated climate, like real climate, is highly variable, and any defined time period will include the effects of climatic variability as well as climate change. It is assumed that by the end of the model run the ratio of climate change signal to the noise of climatic variability is highest. Rescaling by global average temperature change also allows the effect of different emissions scenarios to be compared, assuming that the spatial pattern of change remains constant.

The second and third scenarios are derived, differently, from output from the HadCM2 model. GG1 was produced in a similar way to the HadCM1 scenario, but this time the spatial pattern was based on the last thirty years of model simulation. The spatial patterns were rescaled to the same gobal temperature change as the HadCM1 scenario. The GGx scenario, however, is different in two ways. First, it is based on the average of four experiments, each with the same change in greenhouse gas concentrations but slightly different starting conditions (The GG1 scenario is based on the first of these four experiments). It is assumed that by taking the average of four repetitions ('ensemble experiments') maximises the signal of climate change relative to climatic variability. The second difference is that the scenario is based on the thirty years centred around 2050, and is not rescaled.

Under each scenario, the temperature rise by the 2050s is generally close to the global average, with greater increases in the south and east (exaggerating south–north contrasts). Figure 1.3 shows the change in average annual and summer (June, July and August) rainfall, expressed as percentage changes from the long-term (1961–90) mean.

The percentage change in annual rainfall across the UK is quite different under the three scenarios. The driest conditions are shown under the HadCM2-GGx scenario, with small reductions in the south-east corner of the country. However, this scenario also shows the greatest range in changes with north and east seeing increases in annual rainfall of more 10% in places. The HadCM1 and the HadCM2-GG1 scenarios show increases across the whole of the country, although these are considerably higher in the latter scenario. The consistent patterns in each of the scenarios are the north–south gradient in the changes and the higher increases on the north west coast and over the hills.

The summer scenarios in Figure 1.3 are quite different from the annual scenarios. Most of the UK sees a reduction in summer rainfall, decreasing in the south and east by more than 20% under the HadCM2-GGx scenario. There are some increases, most notably in the far north under the HadCM2-GG1 scenario. Like the annual scenarios, the patterns of change are similar, but the magnitudes vary. Under all the scenarios, winter rainfall increases across the whole of the UK; the increase can be up to 20%.

Higher temperatures generally mean increased potential evaporation. This is exaggerated further by increases in net radiation during summer, but offset in many parts of the UK by increases in relative humidity. Under the HadCM1 and HadCM2-GG1 scenario, potential evaporation may decrease in some upland parts of the north. Increases in annual potential evaporation in the south and east range from around 5–6% under HadCM2-GG1 to around 10% under HadCM2-GGx and close to 20% under HadCM1.

1.4   Implications for the UK rivers and aquifers

1.4.1   Estimating the effects of climate change on hydrological systems

It is important to distinguish between the effects and impacts of climate change. An effect of climate change is a change in streamflow, recharge or water quality; an impact is the consequence of these changes for users of water (including the environment). This section focuses on the hydrological effects of climate change, and the next outlines briefly some of the potential impacts of these changes.

Many studies have explored the potential effects of climate change for river flows, in the UK and elsewhere (e.g. Arnell, 1996a; Arnell and Reynard, 1996; Boorman and Sefton, 1997; Sefton and Boorman, 1997; Arnell et al., 1997; Holt and Jones, 1996), but very few studies have looked at either groundwater recharge or water quality. A reasonably robust methodology for

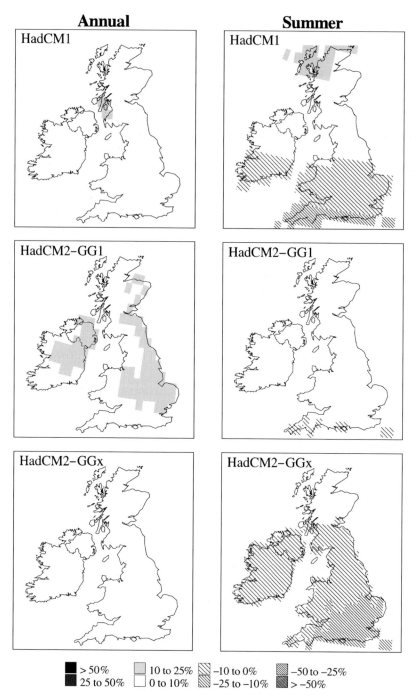

Figure 1.3 Change in average annual and summer rainfall across the UK by the 2050s.

impact assessment has been developed, which involves the definition of climate change scenarios, the perturbation of existing catchment-scale meteorological data, and running a hydrological model with the unperturbed and perturbed meteorological inputs. The biggest uncertainties in this process lie in the definition of climate change scenarios. The previous section summarised the uncertainties in the simulation of the effects of increasing emissions of greenhouse gases on regional climate. When simulating the effects of climate change on hydrology, there are several further uncertainties: (1) downscaling from the climate model resolution to the catchment scale; (2) defining input data at the appropriate temporal scale, and (3) translating the climatic inputs into hydrological outputs.

The first additional uncertainty has attracted considerable attention, and a range of techniques have been developed for downscaling from the climate model resolution (currently around 80 000 km^2 over the UK) to the catchment scale. These methods range from simple interpolation, through correlations between regional and point weather (based on observed data), to the use of weather types to translate regional climate to local weather, and the use of nested regional climate models operating at a higher spatial resolution over a smaller geographic domain (Wilby and Wigley, 1997). These approaches all assume that the broad pattern of climate as simulated by the global climate model is reasonably accurate. Given the uncertainties in the simulation of regional climate, it is perhaps currently not justifiable to use excessively complicated downscaling techniques, although as climate model simulations improve these will become increasingly attractive. The scenarios used in this chapter (Figure 1.3) were downscaled simply by interpolating to a resolution of $0.5 \times 0.5°$.

The second additional uncertainty lies in the creation of scenarios at the appropriate temporal resolution for hydrological analysis, and which represent feasible changes in temporal structure at different time scales. Although climate models operate at sub-hourly time steps, this information is of limited value because it represents behaviour averaged over very large space scales. Most climate change scenarios therefore define changes in monthly climate, and these must be applied to daily – or finer – observed data in order to simulate accurately hydrological processes. The simplest approach is simply to perturb an historical daily time series, but this preserves the present temporal structure. A more complicated approach, but one which has the potential for changing temporal structures, is to use a stochastic weather generator with parameters which can be perturbed given the information contained in a climate change scenario. Current climate change scenarios also tend to define just changes in average climate. Hydrologists and water managers are also interested in changes in the variability in climate over time. Those concerned with floods need scenarios for changes in the frequency of intense rainfall (over various durations), and these changes may not be directly related to changes in average rainfall. Those concerned with low flows are more interested in changes in lower-frequency variability, and specifically with changes in the frequency of multi-month or even multi-year dry periods. Credible scenarios for changes in relative variability cannot currently be derived from climate model output.

The third additional uncertainty lies in the model used to translate climate into hydrological response. Models to simulate moderately high, average and low flows from climatic inputs are currently well developed, although there are some deficiencies in large catchments and catchments with significant groundwater contributions. Models to simulate flood flows are less well developed, and are largely constrained by the spatial and temporal resolution of the available input rainfall data. Groundwater recharge models are poorly developed, and there are relatively few physically-realistic models which can simulate the effects of changes in climatic variables on water quality.

The standard impact assessment methodology implicitly assumes that, in the absence of climate change due to global warming, future climate can be approximated by the present or recent past. This is also a standard assumption in conventional hydrological and water resources assessments, but the previous section has indicated that the UK weather is characterised by multi-decadal climatic variability; even in the absence of global warming, future climate may be rather different from the present (Hulme et al., 1999).

Taken together, these additional layers in the cascade of uncertainty mean that estimates of the effects of climate change are, and will be for several years, highly uncertain. The bulk of this uncertainty lies in the definition of climate change scenarios, ranging from uncertainty in the emission of greenhouse gases to uncertainties inherent in downscaling. In many areas of hydrology, little additional uncertainty is introduced by the hydrological model, although this is not the case with floods, groundwater recharge (to a lesser extent), and water quality.

In quantitative terms, therefore, the effects of climate change may be difficult to determine for a given catchment. However, it is clear that climate change has the potential to alter many aspects of the hydrological cycle. Figure 1.4 (Arnell, 1996b) shows a systems diagram indicating how a change in greenhouse gas concentrations can affect catchment rainfall, evaporation and streamflow, both directly and indirectly through changes in catchment vegetation. These effects may be very complicated, and interactions are important. Higher temperatures, for example, can be expected to lead to increased potential evaporation (because warm air can hold more water), but this effect might be offset by a reduction in net radiation (meaning less energy to evaporate water) or increased humidity. Also, a lack of rainfall may mean that whilst potential evaporation increases, actual evaporation falls.

Catchment land cover can have a significant effect on the water balance and water quality. Both the vegetation type and properties (length of growing season, time of peak growth, plant water use, plant abundance, etc.) can be expected to change with global warming, although the mix of land use

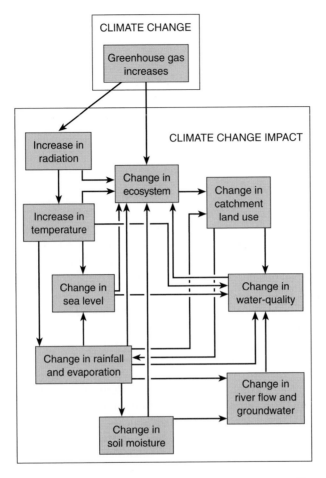

Figure 1.4 Climate change and hydrological processes (Arnell, 1996b).

within a catchment over the next few years is likely to be much more directly affected by changes in agricultural and other land use practices.

Figure 1.4 illustrates how the effects of climate change may be very difficult to define precisely, as they depend on interactions between the different components of the hydrological system, and also how they relate to other changes which may be going on in the catchment. The next three sections look in turn at changes in river flows, groundwater recharge and water quality.

1.4.2 *Climate change and river flows*

As indicated above, there have been several model-based assessments of the effects of climate change on the UK streamflows. Rather than review each

15

Figure 1.5 Locations of six study catchments.

study, this section focuses on three scenarios derived for the 2050s from the HadCM1 and HadCM2 experiments (Table 1.1), as applied to the water balance models used by in a series of studies by Arnell and co-workers (Arnell, 1996a; Arnell and Reynard, 1996; Arnell *et al.*, 1997). Illustrative results are shown for six representative catchments (Figure 1.5).

Change in annual average runoff

Figure 1.6 shows the change in thirty-year mean annual runoff by 2050 across the UK under the three scenarios, calculated using a daily water balance model applied individually to each $0.5 \times 0.5°$ grid cell covering the UK. Under the HadCM1 and GGx scenarios, average annual runoff decreases in southern and eastern UK, and increases in the north and west. Changes can be as high as plus 25% or −20%, with the largest reductions in the south east. Under the GG1 scenario, average annual runoff increases across the whole of the UK.

Change in monthly flow regimes

Both climate change and the water balance vary through the year, so the effects of climate change exhibit seasonal variation. Figure 1.7 shows the

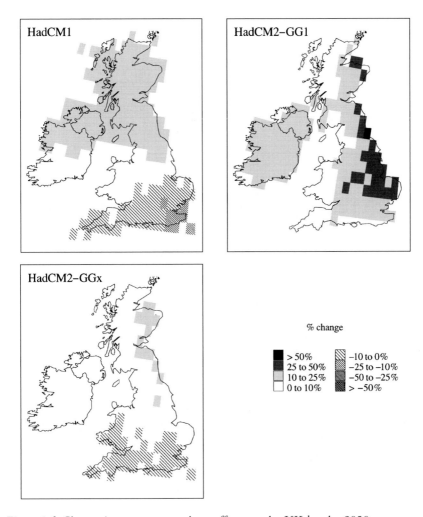

Figure 1.6 Change in average annual runoff across the UK by the 2050s.

percentage change in average monthly runoff for the six study catchments, under the three scenarios. It is clear that climate change will alter the relative variability of flows through the year, particularly in southern catchments. Here, flows increase during winter, but decrease during summer, increasing the range in flows through the year; this happens under all three scenarios. The percentage change in monthly flows can be substantially greater than the change in annual runoff.

Snowfall and snowmelt generally plays a minor role in the UK rivers so, unlike in many cooler environments, there are no major changes in the timing of flows through the year with global warming. A small effect can be

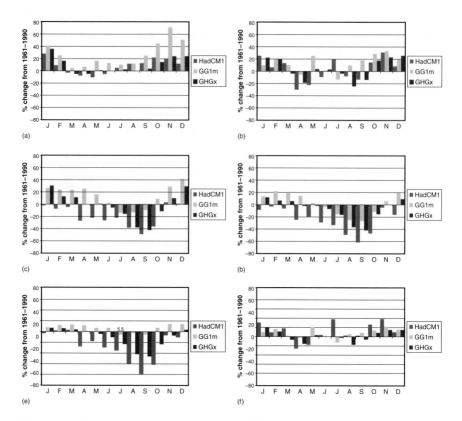

Figure 1.7 Change in average monthly runoff by the 2050s for the six catchments:
(a) Don; (b) Greta; (c) Harper's Brook; (d) Medway; (e) Tamar; (f) Nith.

seen in the upland example catchments (such as the Greta and the Nith),
however, where global warming means a significant reduction in snow cover
and the elimination of the present snowmelt peak. The peak spring flows are
slightly reduced and winter flows further enhanced.

Change in low flow frequencies

The flow duration curve characterises the range in flow magnitudes over
time in a catchment. When the flows are normalised – expressed as a percentage of average flow, for example – then the steeper the slope of the flow
duration curve, the more variable the flows. Flow duration curves under
climate change (except GG1) generally become steeper, particularly in the
south and east, indicating greater variability in flow. The flow exceeded
95% of the time (Q95) is frequently used as an index of low flows. Table 1.2
shows the percentage change in the magnitude of Q95, together with the

Table 1.2 Change in the frequency of low flows by the 2050s.

Catchment	% change in Q95			Average annual days below current Q95		
	HadCM1	GG1	GGx	HadCM1	GG1	GGx
Don	6	16	2	15	11	17
Greta	−5	4	−9	20	17	21
Harper's Brook	−22	−5	−24	34	22	36
Medway	−34	−21	−30	37	28	34
Tamar	−39	−3	−31	41	20	34
Nith	7	3	−4	16	17	19

average number of days per year that flows fall below the current Q95 under the three scenarios (the current average is 18 days per year). Under each scenario, low flows in southern catchments are lower, and current critical values are surpassed more frequently. In the Harper's Brook, Medway and Tamar catchments, for example, flows would be below the current Q95 for around twice as long by the 2050s under the GGx and HadCM1 scenarios.

Change in high flow frequencies

A change in average rainfall due to climate change implies a change in extreme rainfall events, but in actuality the change in storm rainfall magnitudes may not be closely related to the change in the average. Peak intensities may decrease even if long-term averages increase − because rain falls more frequently − or conversely may increase whilst averages decrease, because rain falls less frequently. Scenarios which define changes in monthly rainfall are therefore not so useful for flood assessments as for low flow assessments. Unfortunately, it is currently difficult to derive scenarios for changes in short-term rainfall properties from climate model output, although there are indications that rainfall intensities in mid-latitudes will tend to increase (IPCC, 1996).

Assuming that short-term rainfall changes in line with monthly mean rainfall, it is possible to estimate crudely changes in moderately high flows. Table 1.3 shows the percentage change in the flow currently exceeded 5% of the time, together with the number of days per year on which this flow is exceeded. Under all three scenarios, the magnitude of Q5 and the frequency with which the current Q5 is exceeded are increased in all catchments.

A better indication of changes in flood flows can be obtained by assuming changes in short-term rainfall intensities, consistent with the changes which may result from global warming. Reynard *et al.* (1998) applied a gridded continuous flow simulation model to simulate flood flows in the Severn and Thames catchments (both with an area of around 10 000 km^2), and explored

Table 1.3 Change in the frequency of high flows by the 2050s.

Catchment	% change in Q5			Average annual days above current Q5		
	HadCM1	GG1	GGx	HadCM1	GG1	GG
Don	14	44	14	25	40	25
Greta	13	15	11	23	25	22
Harper's Brook	13	43	23	22	30	24
Medway	3	22	8	19	25	20
Tamar	8	10	3	21	23	19
Nith	13	9	4	24	22	20

Table 1.4 Percentage change in the magnitude of peak floods in the Severn and Thames catchments by the 2050s.

		Return period				
		2-year	5-year	10-year	20-year	50-year
Thames						
	GGx-x	10	12	13	14	15
	GGx-r	6	5	5	4	3
	GGx-s	12	13	14	15	16
	GGx-x + land use	11	12	13	14	16
Severn						
	GGx-x	13	15	16	17	20
	GGx-r	6	6	7	7	7
	GGx-s	15	17	18	19	21
	GGx-x + land use	13	15	17	18	20

Note: GGx-x = GGx scenario with proportional change in rainfall; GGx-r = GGx scenario with change in rain-days; GGx-s = GGx scenario with change in storm rainfall only; GGx-x + land use = GGx scenario with proportional change in rainfall plus 'realistic' land use change.

the effect of several climate change scenarios including GGx. They altered the baseline daily rainfall data in three ways (in each case preserving the same change in monthly rainfall): firstly, by changing every day by the monthly percentage change; secondly, by altering the number of days on which rain falls; and, thirdly, by increasing (or decreasing) rainfall only on those days with rainfall above a certain threshold. They also explored the effect of changing land use. Table 1.4 summarises the change in the frequency of floods with different return periods in the Severn and Thames catchments, under the different variations on the GGx scenario.

These results show that an increase in winter rainfall would lead to increases in flood magnitudes, with the smallest additional effect where the increase is due to more frequent rainfall. They also indicate that the percentage

Table 1.5 Percentage change in average annual recharge by 2050.

Aquifer	Comments	HadCM1	GG1	GGx
Chalk	Yorkshire	5	16	6
Chalk	Berkshire	−9	15	−1
Chalk	Wiltshire	−3	25	3

effect of climate change increases with return period, and that climate change has a much greater effect than realistic land use change in the Severn and Thames catchments. The changes in flood peaks can be expected to be larger in smaller catchments, which are more sensitive to changes in short-duration rainfall, and also the relative effects of land use change can be expected to be larger in smaller catchments.

1.4.2 Climate change, groundwater recharge and groundwater levels

The vast majority of groundwater recharge in the UK occurs during the winter season, once summer soil moisture deficits have been eliminated and before evaporation rises above rainfall in spring. A reduction in summer rainfall and increase in evaporation are together likely to mean longer-lasting soil moisture deficits, so the recharge season can be expected to be reduced. The effects of climate change on recharge will depend on the extent to which this reduction is offset by increased recharge during wetter winters. Also, experience over the last few years has shown that high rainfall alone does not necessarily imply high recharge, as rain that falls in intense events may contribute rapidly to streamflow rather than infiltrate to groundwater (Price, 1998).

Despite the importance of groundwater as a resource in the UK, there have been very few investigations into the potential implications of climate change for recharge. This perhaps reflects a general lack of physically-based groundwater recharge models, which simulate explicitly the generation of recharge from rainfall, accounting for the effects of rainfall intensity. Table 1.5 shows the estimated effect on average annual recharge of the three climate change scenarios for three example aquifers. The three example aquifers differ in their current climate. The estimates were made using a very simple aquifer water balance model as described in Arnell *et al.* (1997). Annual recharge is reduced in southern UK under the HadCM1 scenario, but increased under GG1 and little affected under GGx.

The effect of a given change in recharge on groundwater levels, and hence on discharge to groundwater-dominated rivers, depends on aquifer characteristics. Many chalk aquifers tend to respond rapidly to changes in recharge, with groundwater levels lagging only a few weeks or months behind recharge. These aquifers also tend to have relatively large variations in groundwater level both through the year and from year to year. Permo-Triassic sandstone aquifers, on the other hand, have a much slower response to recharge and vary much less over time. Chalk aquifers are therefore more sensitive to climate change, with groundwater levels showing an earlier response than in Permo-Triassic sandstone aquifers (Cooper *et al.*, 1995).

Changes in river flows in groundwater-dominated rivers are largely dependent on changes in recharge. Changes in summer rainfall will have little effect on flows, and both the total volume of discharge and the seasonal variation will be dependent on winter recharge changes. It is conceivable that higher winter recharge in some catchments could sustain higher flows during summer, even though summers become warmer and drier.

1.4.3 *Climate change and water quality*

Water quality is notoriously difficult to define, but is generally taken to be a function of a river's chemical and biological characteristics. These characteristics are themselves dependent on the volume of water, water temperature, flow generation processes and the characteristics of geology, soil and land use in the catchment. All these properties are potentially affected by climate change.

Water temperature affects the rate of operation of biological and chemical processes within the soil and, more significantly, the river channel. In general, a rise in water temperature will increase the rate of operation of these processes, but different processes are differently sensitive to temperature. Denitrification processes, for example, are more sensitive to a rise in temperature than nitrification processes so, other things being equal, warmer temperatures will mean a reduction in nitrate concentrations. Warmer water can hold less oxygen, so dissolved oxygen concentrations will fall, and also the potential for oxygen-consuming algal blooms will increase. Webb (1992) showed that water temperatures will increase with global warming, at a rate slightly lower than that of air temperature.

Flow generation processes define the way in which water moves to the river channel, and influence how chemical species are flushed from land to river. There have been no published studies into possible changes in the mechanisms of flow generation, but several potential implications can be identified. If soils become dry and cracked, then water will more easily and rapidly infiltrate and move through soil, perhaps transporting more material from the soil and land surface; this tendency would be enhanced if rainfall becomes more intense. Increased flushing, coupled with increased mineralisation of organic nitrogen in drier, warmer soils may lead to increased nitrate peaks in autumn, for example. Conversely, an increase in soil saturation would lead to altered flow pathways and perhaps different rates of transport of chemical species from the soil.

Flow volume has a very significant effect on water quality through dilution. An increase in flow volumes may offset some adverse effects of higher water temperatures, but may exaggerate them further. Low flows during recent droughts have been associated with very high concentrations of particular chemical species (such as nitrates), and very low oxygen concentrations.

Finally, land use has a very significant effect on stream water quality, with agricultural practices being particularly important. Changes in these practices, and water management policies, over the next few decades may have a substantially greater effect on water quality than changes in flows and water temperature. Many of these changes will be independent of climate change, but it is possible that climate change might lead directly or indirectly (through the agricultural system) to land use change.

There have been very few investigations into the effects of climate change on water quality, largely because such studies need complex mechanistic models which incorporate all the appropriate feedbacks and interrelationships. Jenkins *et al.* (1993) made initial assessments of the effects of climate change on nitrate, aluminium and dissolved oxygen concentrations in a number of the UK rivers. The main conclusion, apart from the importance of the relative effect of the different controls, was that water quality in lowland UK rivers – which tend to be less turbulent and more heavily polluted – would be generally adversely affected, whilst water quality in upland rivers would be less impacted. In the Monachyle catchment in upland Scotland, an increase in precipitation and evaporation due to climate change were simulated to lead to increased stream acidification, although the effects were considerably smaller than afforestation (Ferrier *et al.*, 1992).

1.4.4 Conclusions: effects of climate change on hydrological characteristics

This section has reviewed some of the major implications of climate change for river flows, groundwater and water quality. Three general conclusions can be drawn. Firstly, there is a wide range in the amount of research undertaken: changes in annual and monthly flows have been frequently assessed, whilst changes in floods, groundwater recharge and water quality in particular have been less well investigated. Secondly, there is considerable variability between different climate change scenarios. Thirdly, there is a general tendency, however, towards indications of a reduction in streamflows in southern and eastern UK and an increase in the north, with an increase in the range in flows with summer flows considerably reduced over many areas.

The limitations of estimates of the effects of climate change on hydrology are obvious, and relate not just to the scenarios used but also to the mechanistic models used to translate climate into response. Rather then re-iterate these limitations, it is perhaps more important to draw attention to just two: scenarios as currently defined do not incorporate potential changes in year to year variability, and an approach which uses just one baseline period understates the implications of climatic variability for future climates.

1.5   Implications for water resources and their management

1.5.1   Introduction: some general issues

The translation of the effects of climate change into impacts depends critically on both the characteristics of the water management system and how it is managed. Before examining implications for water resources and their management, it is important to emphasise three general points. First, the effects of climate change on hydrology are very uncertain: there is no one 'best guess'. Second, the effects of climate change may be small relative to year-on-year, or decade to decade, 'natural' variability, over the short and medium terms used in much water management. Finally, water resources and their management are subject to a great many other pressures, including changes in demand, legislation and priorities, which may be more significant than changes in the hydrological resource base.

However, the implications of climate change for water management have attracted considerable attention during the 1990s. The 1996 Agenda for Action on water resources (DoE, 1996), initiated after the water resources crises of the early 1990s, urged water managers and regulators to assess the implications of climate change for water resource reliability. In 1999 water companies presented their investment plans to OFWAT, and for the first time these plans took account of the potential consquences of climate change (although whether investment simply to meet the uncertain threat of climate change will be approved by OFWAT remains to be seen). Individual water companies are re-activating dormant plans for new reservoirs.

1.5.2   Potential impacts on water resources

The precise impacts of climate change on water resources depend heavily on the degree of change and the characteristics of the water management system. Table 1.6 summarises (from Arnell, 1998) the major potential impacts, given the climate changes that might occur in the UK. In different catchments and management units, the magnitude and relative importance of these impacts will differ, and not all adverse impacts will necessarily happen in any one catchment. Most of these potential impacts are self-explanatory and require little further amplification, but it is necessary to expand upon a few.

The impacts of climate change on water resources are particularly complicated, as they reflect the effects on the resource base, the demand for the resource and the means to get the resource from source to consumer. Different supply sources are differently affected by climate change, with direct abstractions from unregulated rivers the most exposed. Higher flows during winter may appear to offer increased supplies during winter, but in practice most reservoirs are designed to be full over the winter period and little extra can be stored; the extra flows may therefore not benefit summer supplies.

Table 1.6 A summary of the potential impacts of climate change on water resources in the UK.

Sector	Impact
Water resources	
i. water supply	Reduction in yields, either in total or at certain times of the year
ii. water demand	Increase in demands leading to increase in average and peak requirements. Increased pressure on treatment and distribution system
Flood management	Increase in riverine flood risk, leading to reduction in safety standards
	Increase in urban storm flood risk
Water quality management	Lowered water quality in lowland rivers, with implications for instream ecosystems and water abstractions
	Altered potential for polluting incidents
	Increased potential for combined sewer overflows
Navigation	Lower summer flows leading to reduced navigation opportunities in rivers and canals
Aquatic ecosystems	Altered habitat potential, with species at their environmental margins being most affected
Water-based recreation	Impacts through changes in river flows and water quality

More extreme still, the increased potential for flooding may mean that some reservoirs need to be kept less full in order to provide increased flood storage to protect downstream communities.

Higher temperatures can be expected to lead to increased demands for water, particularly from irrigators and from domestic consumers (primarily for garden watering). With no change in water pricing policies, Herrington (1996) predicted an increase in average domestic demands by 2021 of around 4%, over and above that expected for economic and demographic reasons, with greater percentage increases in peak demands. These increases put pressure on supplies during dry periods and, more particularly, on the capacity of the water supply companies to treat and distribute water during peak periods.

One of the most significant impacts of climate change is likely to be on flooding in urban areas, as higher intensity rainfall events produce flows which exceed the capacity of storm drains. Where these storm drains also carry raw sewage, as in many older urban areas, this increases also the risk of highly unpleasant sewer flooding.

It is important to remember that these climate changes will be super-imposed on the effects of other changes. In the UK, as in many other countries, there is an increasing tendency towards demand management, and measures to bring down demand will lessen the sensitivity of a water supply system to climate change. The more widespread use of water meters, for example, would probably offset the 4% increase in demand due to climate change. On the other hand, increased demands for environmental protection have the potential for reducing available resources, and therefore increasing susceptibility to climate change.

1.5.3 *Implications for water management*

Climate change poses some unusual challenges for water management. On the one hand, estimates of what might happen, and over what time scales, are currently very uncertain, and this uncertainty is unlikely to reduce significantly in the future. This makes it very difficult to develop detailed strategies. On the other hand, water managers can no longer assume that future climate patterns will resemble those of the recent past. Also, different aspects of water management need to consider climate change differently, because the time scales over which they work vary. Actions planned, implemented and changeable over short time scales (less than five years) can ignore climate change, but actions with longer-term implications – such as building large reservoirs or urban sewer systems – need to take climate change into account as they will be difficult to alter in the future.

There are therefore two broad implications of climate change for water management. Because there is no one forecast of the future, water management needs to take a scenario-based approach and consider the implications of alternative futures. These scenarios would not just cover climate change. Scenario-based management is used in major water investments in the United States (Major and Frederick, 1997), and is increasingly being used in business (Schwartz, 1997). The second implication is that adaptive, flexible management responses are preferable in the face of uncertainty to rigid plans or structures which cannot easily be changed.

In general terms, there is a hierarchy of possible responses to climate change, for those parts of water management sensitive to change, ranging from demand-side management, through changes to water management operations, to changes to water management infrastructure. Demand-side management obviously includes restricting the growth in demand for water, but can also be seen to cover changing expectations – by lowering flood defence standards, for example. Operational changes include alterations to the way systems are managed (as happens during drought), and can include legislative and institutional changes (introducing time-limited consents for abstraction and discharge, for example, which can be revised if conditions change). Infrastructural changes include the construction of new flood defences and new reservoirs, and can be either local (catchment scale), regional (water company scale) or strategic (national scale).

1.6 Climate change, hydrology and water resources

This chapter has explored the implications of climate change in the UK for hydrological regimes, water resources and their management. With every 'unusual' event that occurs, interest in and concern over climate change increases, and the water industry is increasingly taking the threat of climate change seriously. Coupled with other changes taking place, climate change has the potential to significantly adversely impact upon the water resource base in parts of the UK; droughts and low flows may become more wide-spread, floods during winter may become more frequent, and water quality is likely to be worsened, at least in lowland rivers.

However, whilst there is growing consensus that climate change is occurring at the global scale, estimates of the effects of climate change on the UK climate over the next few decades are currently very uncertain. There are indications that conditions will be wetter in northern UK but drier in the south, particularly during summer, but different scenarios give very different numerical estimates. Climate change scenarios do not currently define possible changes in year-to-year variability, which may have very significant implications for water management (an increase in the frequency of consecutive dry winters would be very significant). Also, the effects of climate change on streamflow in the UK are likely to be indistinguishable from those of multi-decadal climatic variability over the short to medium term.

What is certain, however, is that the recent past is not necessarily the most appropriate guide to hydrological regimes and water resources over the next few years and decades. Estimates of possible futures can be based on climate change scenarios, but very useful information on the performance of water management systems and potential adaptive strategies can be gained through the analysis of past experience and past hydrological data.

Acknowledgements

The Hadley Centre climate model output used for the creation of the climate change scenarios was provided by Dr David Viner and Dr Elaine Barrow through the Climate Impacts LINK project at the Climatic Research Unit, University of East Anglia.

References

Arnell, N.W. (1996a) *Global Warming, River Flows and Water Resources*. Wiley, Chichester.

Arnell, N.W. (1996b) Hydrology and climate change, in Petts, G.E. and Calow, P. (eds), *River Flows and Channel Forms*. Blackwell, Oxford. pp. 243–56.

Arnell, N.W. (1998) The impact of climate change on water resources in Britain. *Clim. Change*, **39**, 83–110.

Arnell, N.W. and Reynard, N.S. (1996) The effects of climate change due to global warming on river flows in Great Britain. *J. Hydrol.*, **183**, 397–424.

Arnell, N.W., Reynard, N.S., King, R., Prudhomme, C. and Branson, J. (1997) *Effects of climate change on river flows and groundwater recharge*. UKWIR/Environment Agency. Report 97/CL/04/1.

Boorman, D.B. and Sefton, C.E. (1997) Recognising the uncertainty in the quantification of the effects of climate change on hydrological response. *Clim. Change*, **35**, 415–34.

Climate Change Impacts Review Group (1996) *The Potential Effects of Climate Change in the United Kingdom*. Department of the Environment. HMSO, London.

Cooper, D.M., Wilkinson, W.B. and Arnell, N.W. (1995) The effect of climate change on aquifer storage and river baseflow. *Hydrol. Sci. J.*, **40**, 615–31.

Department of the Environment (1996) *Water Resources: Agenda for Action*. The Stationery Office, London.

Ferrier, R.C., Whitehead, P.G. and Miller, J.D. (1992) Potential impacts of afforestation and climate change on the stream water chemistry of the Monachyle catchment. *J. Hydrol.*, **145**, 453–66.

Fraedrich, K. (1994) An ENSO impact on Europe: a review, *Tellus A*, **46**, 541–52.

Fraedrich, K. and Müller, H. (1992) Climate anomalies in Europe associated with ENSO extremes. *Int. J. Climatol.*, **12**, 25–32.

Hadley Centre (1997) *Climate Change and its Impacts: a Global Perspective*. Department of the Environment, Transport and the Regions/Met. Office, Hadley Centre, Bracknell.

Herrington, P. (1996) *Climate Change and the Demand for Water*. HMSO, London.

Holt, C.P. and Jones, J.A.A. (1996) Equilibrium and transient global warming scenarios: implications for water resources in England and Wales. *Water Resources Bull.* **32**, 711–21.

Hulme, M. (1996) *The 1996 CCIRG Scenario of Changing Climate and Sea Level for the United Kingdom*. Climate Impacts LINK Technical Note 7. Climatic Research Unit, University of East Anglia, Norwich.

Hulme, M. and Barrow, E. (eds) (1997) *Climates of the British Isles*. Routledge, London.

Hulme, M., Conway, D., Jones, P.D., Jiang, T., Barrow, E.M. and Turney, C. (1995) Construction of a 1961–1990 European climatology for climate change modelling and impact applications. *Int. J. Climatol.*, **15**, 1333–63.

Hulme, M., Barrow, E., Arnell, N.W., *et al.* (1999) Impacts of anthropogenic climate change in relation to multi-decadal natural climatic variability. *Nature*, **397**, 688–91.

Hurrell, J. (1995) Decadal trends in the North Atlantic Oscillation: regional temperature and precipitation. *Science*, **269**, 676–9.

Intergovernmental Panel on *Climate Change (1992) Climate Change 1992. The Supplementary Report to the IPCC Scientific Assessment*. Houghton, J.T., Callander, B.A. and Varney, S.K. (eds). Cambridge University Press, Cambridge.

Intergovernmental Panel on Climate Change (1996) *Climate Change 1995. The Science of Climate Change*. Houghton, J.T., Meira-Filho, L.G., Callander, B.A., *et al.* Cambridge University Press, Cambridge.

Jenkins, A., McCartney, M.P. and Sefton, C. (1993) *Impacts of Climate Change on River Water Quality in the United Kingdom*. Institute of Hydrology, Wallingford. Report to the Department of the Environment.

Johns, T.C., Carnell, R.E., Crossley, J.F. *et al.* (1997) The second Hadley Centre coupled ocean-atmosphere GCM: model description, spinup and validation. *Clim. Dyn.*, **13**, 103–34.

Major, D.C. and Frederick, K.D. (1997) Water resources planning and climate change assessment methods. *Clim. Change*, **37**, 25–40.

Manley, G. (1974) Central England temperatures: monthly means 1659 to 1973. *Quart. J. R. Met. Soc.*, **100**, 389–405.

Marsh, T.J. (1996) The 1995 UK drought – a signal of climatic instability? *Proc. Inst. Civ. Eng. Water, Maritime Energy*, **118**, 189–95.

Murphy, J.M. and Mitchell, J.F.B. (1994) Transient response of the Hadley Centre coupled ocean–atmosphere model to increasing carbon dioxide. Part II: Spatial and temporal structure of response. *J. Clim.*, **8**, 57–80.

Parker, D., Legg, T.P. and Folland, C. (1992) A daily Central England Temperature series, 1772–1991. *Int. J. Climatol.*, **12**, 317–42.

Price, M. 1998, Water storage and climate change in Great Britain: the role of groundwater. *Proc. Inst. Civ. Eng. Water, Maritime Energy*, **130**, 42–50.

Reynard, N.S., Prudhomme, C. and Crooks, S. (1998) Impact of climate change on the flood characteristics of the Thames and Severn rivers, in *Proceeding 2nd International RIBAMOD Conference*. HR, Wallingford, February 1998. Institute of Hydrology, Wallingford.

Schwartz, J. (1997) *The Art of the Long Term*. Wiley, Chichester.

Sefton, C.E. and Boorman, D.B. (1997) A regional investigation into climate change impacts on UK streamflows. *J. Hydrol.*, **195**, 26–44.

Shorthouse, C.A. and Arnell, N.W. (1997) Spatial and temporal variability in European river flows and the North Atlantic Oscillation, in *FRIEND'97 – Regional Hydrology: Concepts and Models for Sustainable Water Resource Development*. International Association of Hydrological Sciences Publication, **246**, 77–85.

Webb, B.W. (1992) *Climate Change and the Thermal Regime of Rivers*. University of Exeter, Department of Geography. Report to the Department of the Environment.

Wigley, T.M.L. and Raper, S.C. (1992) Implications of revised IPCC emissions scenarios. *Nature*, **357**, 293–300.

Wilby, R. (1993) Evidence of ENSO in the synoptic climatology of the British Isles since 1880. *Weather*, **48**, 234–239.

Wilby, R. and Wigley, T.M.L. (1997) Downscaling general circulation model output: a review of methods and limitations. *Prog. Phys. Geog.*, **21**, 530–48.

2

LAND USE CHANGE

Mark Robinson, John Boardman, Rob Evans, Kate Heppell,
John Packman and Graham Leeks

This chapter considers the changes in land use in the UK and their hydrological impacts. It focuses on agriculture, forestry and urban development
and their implications for hydrology, soil erosion and sediment transport.

2.1 Land use changes in the past and present

Changes in land use, both cover and management, will affect river flows
and the water balance. Vegetation cover and land management have a major
influence on the apportionment of precipitation to evaporation, streamflow
and groundwater. The land use of Britain has been undergoing changes for
thousands of years. Currently about 70% of the UK is agricultural crops
and grassland, whilst 10% is woodland and 10% urban areas (housing and
commercial uses). The remaining 10% includes derelict land and mineral
workings (Barr, 1993; DoE, 1993, 1996).

About 5000 years ago, before the population of the UK was sufficiently
numerous to make a serious impact on the landscape. the land was predominantly covered by woods; tree cover was sparse only where the ground was
too wet or too exposed. Woodland clearance was brought about not only by
axe and fire, but by grazing animals stopping its regeneration. Many areas,
especially lowlands and lower hills, were cleared of their trees more than
once; for example, when the population declined after Romano-British times
and the Black Death in the middle of the fourteenth century, trees recolonised
abandoned land. Up to the present about 97% of the original woodland
cover has been cleared (*Guardian*, 1997).

The pre-historic clearance of woodland from the UK landscape would
have brought about a major change in its hydrology. Thereafter, changes
would probably be related to the changing proportions of the land under
forest, moorland, grass, and crops, and how intensively that land was used,
for example, the number of animals grazing the land, or whether annual
crops were grown.

Since the Second World War, Britain has had a policy of encouraging food production through grant-aid, technical advice and price support, particularly since joining the European Economic Community. However, since the 1980s the Common Agricultural Policy (CAP) has been reformed to some extent, to try to reduce overproduction of certain foodstuffs. There has been a progressive shift to more sustainable and environmentally friendly methods of food production, and the protection of the environment has been integrated into the CAP. This has resulted in a sharp decline in the crop area (from 5.3 Mha in 1987 to 4.5 Mha in 1994), as a result of the European Commission Set-Aside schemes established in 1988 to reduce the amount of arable land. In 1994 there was 0.75 Mha of set-aside in the UK. Between 1983 and 1994 the total area of farm land in the UK fell by 0.25 Mha, whilst forests and urban areas increased.

At present, only about 10% of the UK is wooded, although this is about twice what it was at the beginning of the twentieth century; just over half of this is coniferous woodland. Forest evaporation rates are generally greater than either grassland or arable land, particularly in the high rainfall upland parts of the country, and it is likely that runoff now is significantly greater than it was when the UK was largely wooded. Contemporary comparisons of catchments with different vegetation covers illustrate this point. Forestry also affects, not just water quantities, but also water chemistry (e.g. Kirby et al., 1991; Neal, 1997; Robinson, 1998) and sediment transport (Leeks and Marks, 1997).

The clearance of the original forests would have increased runoff and probably caused erosion of valley floors and headward extension of channels, especially where the land was largely unenclosed and herds of cattle or sheep grazed the valley floors. Wetter soil conditions would have encouraged the growth of peat soils on flat and gently sloping hill tops, and expanded onto lower slopes as forests were cleared. It is often claimed that peat bogs absorb rainfall and reduce flooding, but there is still much debate upon the exact role of wetlands in influencing river flows. Peat bogs eroded in medieval times. This erosion was probably triggered by clearance of the woodland margins and by monastery-owned sheep grazing the moors. Erosion became more widespread after the onset of the Industrial Revolution and its accompanying airborne pollution which killed the bog mosses. This, together with moorland fires, accidental and as part of heather and grouse management schemes, may have led to peat erosion and gullying which would speed the transport of rainfall and sediment from the moors.

Animals grazing the improved pastures in enclosed fields, and lowland heath and upland moor, may compact the soil and promote runoff. Evans (1996) suggests that changes in sheep grazing on moors could help to 'explain' differences in runoff, citing declining runoff in the period 1930–55 in a moorland catchment in North Wales after sheep numbers declined, and increasing runoff in the period 1945–75, in the Derwent catchment in the Peak District as sheep numbers rose. The number of sheep in the hills of the UK has increased markedly since 1945, often by a factor of 1.5 or more (MAFF, various dates). In 1872 there were 17.9 million sheep in England, most of them in the lowlands; this rose to 18.6 million in 1944, and 20.5 million in 1992, most of them in the hills. Numbers peaked in 1990 (20.8 million) and have declined since. In the same years, cattle numbers increased to a peak of 8.3 million in the late 1970s, but have then declined due to government moves to reduce their numbers and the costs of subsidies.

Runoff will be greater if a large proportion of the crops are sown in autumn, leaving the land smooth and almost bare over winter, the time when much of the rainfall runs off the saturated ground. Winter cereals were especially widely grown in Roman times, as they are now, when they cover almost half the arable land. Other crops, including potatoes and sugar beet, for example, when grown on a field scale can also promote runoff. Sugar beet was introduced by government to boost farm incomes in the 1920s. Maize, another crop recently introduced to the UK for economic reasons and rapidly expanding in area, also promotes runoff. Runoff is greater and more rapid from arable cropped land than from closely vegetated permanent grassed slopes. So, under a given climatic condition, runoff in lowland UK is likely to have been greater at times when cropped land was at its greatest extent, generally at times of population pressure. This included the Romano-British period, the thirteenth and fourteenth centuries, the Victorian 'High Farming' period which ended in the mid-1870s, the First and Second World Wars, and since 1945 (Evans, 1992). Arable land may have been even more widespread in Medieval times than now, because the climate was actually warmer than today and agriculture was still at an extensive, subsistence level. Medieval ridge-and-furrow can still be found in pasture high up in the Cheviots and in grass fields on heavy clay soils in the Midlands of England. In 1872 the ratio of land under crops to improved grass in England was 0.42, whereas in the 1930s, at the height of the UK agricultural depression, it was 0.27. During the Second World War, this rose to 0.49 in 1944, and in 1988 prior to cut backs imposed by the Common Agricultural Policy it peaked at 0.53 (MAFF, various dates). However, over this time the total land area under crops and grass declined from about 10 to 8 Mha as land has been built on, which will have further increased runoff.

In parallel with the agricultural expansion in the 1960s and 1970s, just as in the period of 'High Farming' in the nineteenth century, there was a large increase in the area of farmland with artificial in-field drainage; much of this was accompanied by large scale arterial channel improvements. This drainage facilitated a great increase in both the extent and the intensity of agricultural production which included the expansion of cereal crops. Current rates of new and renewal drainage are low, and the overall stock of drained land is slowly declining (Robinson and Armstrong, 1988).

In some areas, particularly where high value crops are grown in eastern England, there has been a rise in the use of irrigation water to ensure crop quality (particularly for potatoes, and also for sugar beet). Agricultural abstractions (of which approximately half go to irrigation) account for only 1% of total water supply in England and Wales (DoE, 1992), and is concentrated in the Anglian and Severn Trent regions. This water is needed during the summer growing season, which is the time of greatest stress on water reserves, either surface or groundwater. Drip irrigation is used on horticultural crops whereas, for arable crops, spray irrigation is used. The latter is particularly wasteful of water if carried out during daylight hours when evaporation losses are greatest. Irrigation is expected to continue to increase as it provides producers with higher quality produce and a greater reliability of yields that are less dependent upon weather conditions. Irrigation water usage is predicted to increase by an annual rate of 1.7% to the year 2001, and then at 1% to 2021 (NRA, 1994).

At the present time, there is a large body of information available about the water use requirements of many crops during their growing season, but in contrast very little is known about their overall annual water use, or the water balance of bare soil, fallow or set-aside.

2.2 Relevance to hydrology

Evaporation rates of water intercepted by plant canopies are known to be higher for forests than for shorter vegetation (Calder, 1992). Plants also supply water to the atmosphere by transpiration, extracting water and nutrients from the soil. Different plants will have different water usage and be affected to different degrees by soil water availability. The fate of precipitation that reaches the ground surface depends critically upon vegetation and soil processes. If the rainfall intensity is greater than the infiltration capacity of the soil (for example due to compaction by farm machinery or impervious surfaces in urban areas) then surface (overland) flow may occur. Water that infiltrates into the soil moves downward if the soil is permeable or may flow laterally (sideways) if there are impeding layers (e.g. a plough pan). Changes to soil structure which may influence water movement include ploughing for crops and ripping to remove impeding layers. Both of these will open structural macropores which may encourage water movement to deeper soil layers. Conversely, seed drilling may result in denser, more compacted soils.

Land use changes, comprising changes in vegetation cover as well as in land management techniques, may have a range of hydrological and water quality impacts. Some of these are reviewed below in more detail.

2.3 Drainage

A major change in recent years in Britain and elsewhere is the installation of field drains in farmland. The UK (and particularly England) is now one of the most extensively drained countries in the world. Tile drainage originated in the UK, and in the 1970s alone, about 10% of the farmland of England and Wales was drained (Robinson and Armstrong, 1988). The benefits of installing artificial drains, comprising systems of plastic or clay pipes, typically 0.75–1 m deep, include improved crop yields, a longer working season when heavy machinery can be used on the land, earlier emergence and ripening of crops and plants, and less animal disease. These drains will lower the water table and soil water content in the fields as the water is removed rapidly to nearby watercourses. This may reduce the potential for recharge from the fields to the regional groundwater table, although the supply of gravity drained water from the upper metre of the soil will help to sustain river flows when normal soil water outflow from undrained land has ceased.

Field drainage in Britain probably dates from Roman times, but it is only in the last 200 years that significant amounts have been carried out. The most primitive technique was to plough the land into large scale ridges and furrows, and open ditches have long been used to drain fields. The only effective land drainage technique which does not disrupt field cultivation is a system of sub-surface pipes. Drains remove water from the saturated zone, below the water table, through the action of gravity. The most common form is regularly spaced parallel lateral drains which feed into a collector drain and then into an arterial drainage channel. Following the invention of machinery to manufacture cheap extruded clay drainage pipes, underdrainage became widespread in the mid- to late-nineteenth century period of 'High Farming', and there is evidence that drainage in the nineteenth century was even greater than that in the twentieth century (Green, 1980; Robinson, 1986). The geographical variations in nineteenth century drainage in England and Wales, based on later ground surveys of the presence of these old drains, shows good general agreement with environmental constraints on agriculture, principally climate and soils. Higher rates of drainage activity occur in areas with more impermeable soils and the highest rates are found in the north and west where there is high rainfall and the soils are of low permeability, including heavy clays.

Many nineteenth century drains reached the end of their useful life following the First World War. Government grant-aid payments towards the cost of drainage were introduced in 1921 in Scotland and in 1939 in England and Wales. These payments encouraged a new phase of drainage and marked the start of relatively reliable records of drainage work carried out (Robinson and Armstrong, 1988). The implementation of field drainage increased steadily to a peak of about 100 000 ha y^{-1} in England and Wales in the 1970s, and since then has fallen to under half that rate due to rising surpluses of agricultural production in the European Union, a reduction in government support and a more uncertain economic future. About one million hectares were drained in the decade 1971–80, about 10% of the agricultural land of England and Wales.

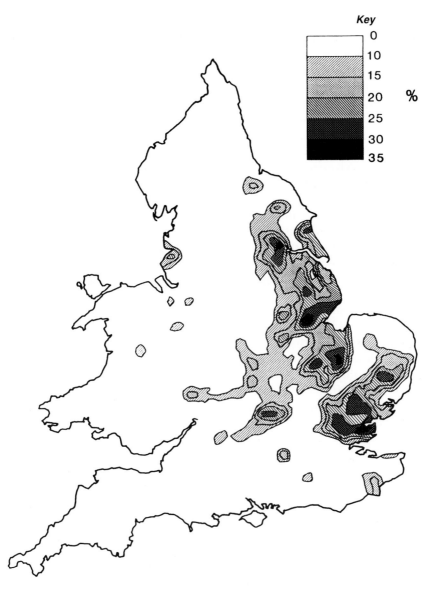

Figure 2.1 Percentage of agricultural land in England and Wales with pipe drainage
installed during the period 1971–1980.

In contrast to the nineteenth century, the distribution of drainage in the
twentieth century reflects mainly economic factors, with the majority of field
drainage taking place in the south and east (Figure 2.1). Arable land there is
dominated by cereal production, autumn access onto the land is vital for the
cultivation of autumn cereals and, wherever soil wetness is a limiting factor

on the working season, drainage has become an essential part of continuous arable cultivation. Apart from the chalk lands, which do not require drainage, there is a rough correspondence between the areas of high drainage activity and the cereal lands. Additional concentrations of drainage activity are in the silt and peat lands of the Fens of East Anglia and the coastal plain of Lancashire where there are important areas of high value horticulture. In these areas, there is a high investment in equipment which must be recovered by high value crops such as maize, sugar beet and potatoes. The most common type of surface water problem is poor drainage on a low permeability soil. Secondary treatment, such as sub-soiling, is frequently used to loosen the sub-soil over the pipe drains and to facilitate the movement of water to the drainage system.

Robinson and Rycroft (1999) examined the impacts of sub-surface drainage and concluded its impact on high flows of rivers was most likely to be increased flooding downstream in areas with moderately permeable soils and extensive field boundary ditches.

In upland, moorland areas such expensive drainage techniques cannot be justified and the main technique adopted is the cutting of open drains typically 40–45 cm deep and with spacings of the order of 20 m. This is frequently carried out for sheep grazing and grouse moors. Records of this 'cheap' drainage are much more limited than those for pipe drainage, but broadly reflect the distribution of the principal controlling factors of soil type and climate.

Land drainage may also affect the water levels in surrounding land and so may pose a threat to nearby wetlands when it occurs in close proximity and if the permeability of the intervening land is sufficiently high to allow an appreciable drawdown to occur. Wetlands may be at risk if neighbouring land is drained, whether for farming, forestry or peat extraction.

2.4 Agricultural soil erosion

Soil erosion is not a new phenomenon in the UK; there is ample evidence of erosion since original forest clearance, largely as a result of subsequent cultivation. The major loss of soil probably occurred in the Bronze and Iron Ages in southern England (Bell, 1992; Favis-Mortlock *et al.*, 1997). Associated with clearance and cultivation, there must have been major changes in run-off from slopes and in the hydrology of rivers; these changes have been little investigated. It is likely that in chalk landscapes water tables were higher and streams more active than in the twentieth century. In the last two millennia, erosion has been episodic and concentrated in periods of population pressure and expansion of arable agriculture (Evans, 1990a). There is little evidence to relate erosion to climate change, but doubtless large storms were occasionally very effective.

Since the Second World War, the expansion of the area under arable crops in the lowlands, and the increases in stock numbers in the uplands, have led to an increase in erosion. Arable crops have spread to steeper slopes, are grown in larger fields and cultivated with heavier machinery. Crop rotation systems have declined in favour of monocultures, less organic manure is used and organic matter levels in soils have decreased which affects aggregate stability. In the south and east, the widespread adoption of winter cereals from the mid-1970s onwards, greatly increased the risk of erosion due to the coincidence of bare ground after drilling and wet weather in October–December. On the South Downs, this is the dominant form of erosion. In areas where fruit, vegetables and maize are grown, there is also a risk of erosion in the early summer months; sandy soils in the Midlands are vulnerable to summer storms.

A monitoring scheme in the mid-1980s recorded rates of erosion in seventeen localities in England and Wales (Evans, 1993). As a result, the soil associations most at risk of erosion are known (Evans, 1990b). The National Soil Map identifies at risk soils, including a small number at risk of wind erosion (Soil Survey of England and Wales, 1983). More recently, the risk of erosion on land under winter cereals has been mapped (Soil Survey and Land Resource Centre, 1993). A ten year study of erosion on the South Downs shows that 92% of erosion occurred in three years with most of that a result of one particularly wet autumn, 1987 (Boardman, 1998a). Monitoring studies in the UK indicate that average rates of erosion are low, generally between 1 and 5 m^3 ha^{-1} y^{-1} (Boardman, 1998b). However, individual fields may lose as much as 200 m^3 ha^{-1} y^{-1} as a result of extreme events (Boardman, 1988). Average rates are low because summer storms tend to be spatially limited and in many areas of the country there is little bare ground in summer; in areas growing winter cereals, the 'window of opportunity' for erosion is about two months before crop growth inhibits runoff and soil loss.

The delivery of soil to streams from sources on agricultural land has been little studied (but see Foster *et al.*, 1990; Walling, 1990; Slattery *et al.*, 1994). Soil tends to be transferred short distances to sites of storage such as field boundaries and flood plains. During extreme events, hillside slopes may be linked to valley-bottom flow (ephemeral gullies). On chalk landscapes lacking permanent stream networks, soil is generally transferred to valley bottom and valley side storage sites and does not reach streams or the sea.

Erosion in the uplands occurs on both mineral and peat soil. High precipitation and steep slopes increase the risk but only where grassland or woodland is disturbed by mass movements, fire, forestry operations or human and animal trampling do erosional features develop. Evans (1997) reviews the effects of animals, mainly sheep and red deer, on erosion.

Impacts of erosion include sedimentation of reservoirs and decline in water quality (Butcher *et al.*, 1993), and damage to property, principally roads and houses. Many such cases are not documented but occur regularly in the Isle of Wight, the East and West Midlands, Somerset and Kent (DoE, 1995; Evans, 1996). On the South Downs, Boardman (1995) lists thirty five sites of damage in some cases many times repeated, in the years 1976–93.

The influence of climate on erosion is sometimes exaggerated. In the 1980s, 'a decade of drought' (Newson, 1989), erosion increased substantially due largely to land use change. Climate change in the future poses some risks. Wetter winters may increase erosion on cereals. Similarly, an increase in the number or intensity of summer storms would pose a risk in some areas and for some crops, e.g. potatoes (Boardman *et al.*, 1990; Evans, 1996). Perhaps of greater significance is the possibility of land-use change driven by climate, e.g. the spread of maize and other high-risk crops into southern Britain. In the uplands, an increase in temperature may encourage conversion of grassland to arable including grass leys.

2.5   Forestry

The UK began the twentieth century with a much depleted forest area, and a reliance on cheap timber imports from the British Empire, such as Canada. Since the First World War, it has been government policy to encourage home-grown timber production through the state Forestry Commission and private forests encouraged by tax benefits and subsidy. Formerly, planting was predominantly of conifer plantations on less valuable farmland in the uplands. However, there has recently been government encouragement of new plantings of indigenous broadleaved species through the Broadleaved Woodlands Grant scheme in 1985 and the Woodland Grant scheme since 1988. More recently, the government established twelve Community Forests and a National Forest in England on mainly derelict, formerly urban and industrial, land and, most important of all in terms of total area, it announced in 1995 its aim to double the area of forest in England (HMSO, 1995), and to increase that in Wales by 50% and encourage further expansion in Scotland and in Northern Ireland. This is potentially the most dramatic land use change for many centuries, both in terms of its extent – over 10 000 km^2 – and the nature of the change from short vegetation to much taller woodland.

Extensive information exists for water use of upland coniferous forests in relatively high rainfall environments (Johnson, 1991) from work carried out by the Institute of Hydrology at Plynlimon in Wales (Kirby *et al.*, 1991), Balquhidder in Scotland, Coalburn on the English/Scottish border (Robinson, 1998) and elsewhere. Annual evaporation 'losses' (i.e. the difference between annual precipitation and annual runoff) in the largely forested Severn catchment amount to about 200 mm more than the grassland Wye catchment (Kirby *et al.*, 1991). This represents a 15% reduction in the flow. Similarly, increases in stream flow result from tree felling (Johnson, 1991).

Forestry is now heavily mechanised (Plate 2.1) and when surface vegetation and soils are disrupted by surface ploughing, track and drain construction, as is commonly the case during upland forestry operations (Newson, 1980), there is the potential to mobilise large quantities of weakly-cohesive glacial and fluvio-glacial material. Once sediment is mobilised, plough furrows,

Plate 2.1 Mechanised tree felling in upland Wales (Steve Marks).

drains, ditches and steep slopes leading to down river headwater channels, can facilitate the delivery of sediment to water courses and increased rates of sediment transport (Newson and Leeks, 1987). Due to the extensive planting that occurred during the twenty years after the Second World War, many British forests are now reaching the felling stage. Enhanced sediment delivery to fluvial systems has been associated with all stages of the forest rotation as quantified by long-term studies of fluvial sediment processes in the Institute of Hydrology Plynlimon experimental catchments in mid-Wales (Leeks and Marks, 1997 (Figure 2.2)). In comparing both bedload and suspended sediment yields from headwater catchments, grassland sediment yields are one-third to one-sixth of those from forests with even higher yields associated with afforestation and felling phases. The impacts of particulate outputs on the ecology of freshwater ecosystems is reviewed by Marks and Rutt (1997). In-channel accumulations of coarse woody debris can have major geomorphological effects through its influence on the storage and transport of particulates and organic matter, channel form and stability. The impacts of enhanced sediment loads associated with afforestation upon water supply in Wales, Scotland and in the north of England have been reported in case studies of increased treatment costs or damage to distribution systems and accelerated deposition within downstream storage areas including reservoirs (Duck, 1985; Stott, 1989; Leeks, 1992; Burt *et al.*, 1984; Stretton, 1984). The range of impacts described above have been widely recognised by forestry operators, conservationists and the water industry (Maitland *et al.*,

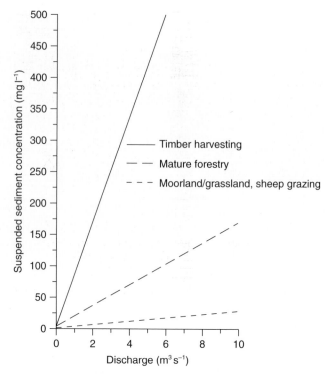

Figure 2.2 Annual (1996) rating curves of suspended sediment concentrations against discharge for headwater catchments affected by different land-use practices within the Institute of Hydrology's Plynlimon Experimental Catchment.

1990) leading to new and modified approaches to forest management (Forestry Commission, 1993).

Much less is known about the water losses from broadleaf trees on lowland areas. Yet this is the species/climate mix targeted by the proposed expansion of forestry. The location of the new woodlands is still under discussion, but present indications are that policy makers favour the lowlands (Forestry Commission/Countryside Commission, 1996). Only a few hydrological studies have been carried out on broadleaved woodlands, but it appears that in drought years forest evaporation losses will be higher than from grassland or arable land, with resultant lower streamflow and reduced recharge (Harding *et al.*, 1992; House of Commons, 1996).

Land use change is heavily dependent upon European Union (EU) agricultural policy affecting the type, area and variety of farm crops. Within the UK, there are a number of policy initiatives within the general reform of the CAP, and perhaps most notably the series of Rural White Papers for England, Wales and Scotland. These highlighted the intention to promote a massive expansion of forested land, partly in response to agricultural surpluses and partly in response to Agenda 21 of the 1992 Rio Earth Conference. The Rural England White Paper (HMSO, 1995) called for a doubling of the area under forest, from about 7 to 14% (an additional area in excess of 9000 km^2) by the year 2045. This is about seven times larger than the total planned planted area for the Community and National Forests. Trees are known to have higher evaporation rates than short crops, both in terms of their interception of rainfall due to their greater height and aerodynamic roughness and their transpiration due to deeper rooting (Calder, 1992). At Thetford in East Anglia, the forest cover reduced groundwater recharge by 50%. There is the potential for serious water resource implications including groundwater and dry weather flows, particularly since much of the proposed new forestry is likely to occur on lowland areas in England, such as the Midlands where there is a high density of domestic and industrial water users. This increased area of broadleaf forest may result in much reduced recharge to underlying aquifers that are major water resource suppliers. In addition to conventional woodland plantations, the White Paper for England also advocated the planting of willow and poplar to be used as short rotation coppice (SRC) as a biofuel. Recent studies of SRC have shown that they have the potential for extremely high transpiration losses, that are even greater than those from open water (Hall and Allen, 1997).

Current climate change scenarios suggest that the southern part of Britain will suffer from drier summers. The resultant water shortages will potentially be magnified further by any change from short vegetation to deeper rooting trees.

2.6 Use of pesticides and fertilisers

2.6.1 Pesticide usage

Herbicides, fungicides, insecticides and wood preservatives are all examples of different types of synthetic pesticide. Since their initial production in the 1940s, the largest growth in pesticide use in the UK has been associated with those fungicides and herbicides which control pests in autumn-grown cereals. By the 1960s, their use had enabled farmers to move from more traditional systems of crop rotation to continuous cereal cropping (Ward et al., 1993). As a result, the weight of pesticide applied to land rose from approximately 15 000 tonnes per annum in 1974 to over 30 000 tonnes per annum in 1982. The last decade has seen further significant changes in patterns of use for both agricultural and non-agricultural purposes. For agricultural pesticides, these changes have been driven mainly by variations in cropping patterns, allied to economic factors such as changes in levels of financial support. The use of non-agricultural pesticides has risen, chiefly through an increase in their application to hard surfaces (roads, railways and car parks) for weed control.

Table 2.1 Top ten pesticides used on all agricultural and horticultural crops throughout Great Britain, ranked in descending order of weight applied in 1994 (t).

Rank (by weight) active substance	Type	Main uses	Weight applied in 1994	Change since 1984 (%)
1. Sulphuric acid	Desiccant	Potatoes	12 997	+31
2. Isoproturon	Herbicide	Wheat, winter barley	2382	+56
3. Chlormequat	Growth regulator	Wheat, winter barley	2335	+110
4. Mancozeb	Fungicide	Potatoes, wheat	1191	+235
5. Sulphur	Fungicide	Wheat, sugar beet, oilseed rape	1113	+75
6. Mecoprop	Herbicide	Wheat, winter barley	957	−79
7. Chlorothalonil	Fungicide	Wheat, pulses	841	+801
8. MCPA	Herbicide	Wheat, winter barley	724	−54
9. Glyphosate	Herbicide	General weed control	606	−30
10. Mecoprop-P	Herbicide	Wheat, winter barley	578	n/a[a]

Source: Pesticide Usage Survey Report 100 (MAFF).
[a] isomer not available in 1984.

Data on pesticide usage can be presented in a number of ways, but the most common and simplest method is to give the weight of the active substance applied (Table 2.1). The weight of agricultural pesticides applied to crops in the UK has fallen by over 7000 tonnes per annum (17%) in the last decade (MAFF, 1997). This can be attributed to:

- the advent of more efficient or selective pesticides which require lower dosage rates, such as mecoprop-P (the active isomer of mecoprop); and
- the adoption of reduced dose application rates because of label recommendations or due to the perception of a diminished threat of pest invasion.

While the weight of dosages has fallen, the frequency with which farmers have applied pesticides has increased. It has now become common practice for a cereal crop to receive six or seven applications of different pesticides in a single growing season (NRA, 1995). As a result, in the period between 1984 and 1994, during which time almost one million hectares of land were taken out of cereal production, there was an apparent increase in the amount of pesticide applied to land as measured by changes in usage by area.

In 1974 the non-agricultural use of pesticides was dominated by those chemicals used as wood preservatives and the herbicide, sodium chlorate, which was used for weed control on uncropped land. Herbicide usage in non-agricultural conditions has steadily increased since then, so that by 1989 an estimated 500 tonnes of active ingredient were used per year for both total weed control of paved areas, and for selective weed control in parks and playing fields. Although this is still a small percentage of the total pesticide use in the UK, it is particularly important in those areas where pesticides are applied to hard surfaces from which rapid runoff to drains and soakaways during rainfall is likely (Heather and Carter, 1996).

2.6.2 Fertiliser usage

The rapid rise in cereal production which occurred from the 1970s onwards, with its associated use of new technologies to raise crop yields, not only increased the use of agrochemicals such as pesticides but also encouraged the increased use of chemical fertiliser products. During the 1950s, approximately 210 000 tonnes of inorganic nitrogen fertiliser were used in the UK per annum. By 1990 this figure had risen to over 1.5 million tonnes per annum (NRA, 1992). Currently, approximately 2 million tonnes of mineral fertiliser nutrients (nitrogen, phosphate and potash) are used annually in the UK (MAFF, 1996). Animal-based agriculture has also intensified and large applications of manure (organic fertiliser) have therefore become increasingly common. The EC Nitrate Directive (91/676), which concerns the protection of waters against pollution caused by nitrate from agricultural sources, recognises that excessive use of fertilisers poses an environmental risk, and suggests that this risk applies to both inorganic and organic (slurry and farmyard manure) fertilisers (Heathwaite et al., 1993).

All fertilisers contain varying percentages of those nutrients that are essential for plant growth: nitrogen, phosphate, potassium (potash), magnesium and sulphur. The exact chemical composition of the fertiliser to be used depends on, amongst other factors, the requirements of the crop to which the fertiliser will be applied. For instance, to a cereal crop such as winter wheat or barley, a farmer might apply an autumn application of a phosphate or potash-based fertiliser, such as Triple Super Phosphate or Muriate of Potash, and a spring application of a nitrogen-based fertiliser such as ammonium nitrate. Organic fertilisers such as slurry and liquid sewage sludge contain appreciable quantities of nitrogen in soluble forms.

Between 1970 and 1985, the application of fertiliser to winter wheat doubled, and to winter barley rose by two-thirds. Since the mid-80s, however, the total nitrogen application for tillage crops in Great Britain has gradually declined to an average value of 145 kg ha^{-1}, as a result of both a drop in average field application rates and a decline in the extent of crop area treated (MAFF, 1996). On grassland, the total nitrogen application rate has declined from over 130 kg ha^{-1} in 1985 to 115 kg ha^{-1} in 1996. This has been accompanied by a reduction in the extent of grassland used for grazing in Britain. The rates of phosphate and potash usage on tillage crops have remained largely unchanged for the last five years (about 69 and 82 kg ha^{-1} respectively in 1996). Average application rates to grassland, however, have declined slightly to approximately 34 kg ha^{-1} for phosphate and 45 kg ha^{-1} for potash in 1996.

2.7 Urban development

According to DETR (1998a), 88% of the population of England live in urban communities (of greater than 1000 people). This may be compared with estimates by the UN Centre for Human Settlements (1996) that almost 50% of the world's population live in urban areas. Notwithstanding differences in definitions and the immense problems in constructing global estimates, these figures make the point that Britain is one of the most heavily urbanised countries. Also according to DETR (1998b), land in urban/industrial use in England is predicted to increase from 10% to just under 12% by 2016, with a need for 4.4 or even 5 million new homes. There is a steady increase in the demand for housing and other urban development, currently of about 15 000 ha per annum. Of this, about half comes from the transfer

Plate 2.2 An extreme example of an urbanised river, River Tame near Birmingham (Mike Acreman).

of formerly agricultural land and the remainder comes from the re-use of previously developed land, reflecting the policy of local and central government to recycle previously developed and derelict land for urban use.

Before the Industrial Revolution, urban areas had long existed as centres of trade, culture and government, but mechanisation of agriculture and industrial development provided 'push' and 'pull' factors that initiated large-scale migration from rural areas into towns and cities. Despite pressures on housing and services, the concentration of economic, technical and social stimuli generally brought greater prosperity and further urban growth as solutions to the problems evolved. These solutions have included high-rise buildings, mass-transport systems, dormitory and satellite towns, and out-of-town commercial parks. Although each solution has brought further problems to resolve, the general process of urbanisation has continued, and the land area covered by streets and houses has continued to grow.

The impacts of such urbanisation on catchment hydrology are great and varied, depending on the local environment, development history, and technical strategies adopted to meet the water needs. These needs include a water supply for domestic and industrial use, a waste water disposal system, a drainage system to control groundwater and remove local flood water, and a flood defence system to protect against inundation from external areas. The urban impact depends on how developed are these 'artificial' systems, and how they interact with the 'natural' processes of rainfall, groundwater and drainage that are themselves modified by urban influences (Figure 2.3).

Consider first the water supply and waste water impacts. A typical scenario might involve an inland city, where a combination of over-abstraction from local surface and groundwater sources with their pollution by poor waste control would require increased and cleaner water supplies to be imported from a remote source. While originally the falling surface and groundwater discharges might have caused local rivers and streams to dry and allowed their courses to be polluted, piped and backfilled, a subsequent reduction in local abstraction, combined with leakage of imported water and garden watering might bring groundwater flooding, remobilisation of soil pollutants, and infiltration into wastewater and storm drainage systems. The imported water supply meanwhile is used in a rhythmic daily pattern, collected in a waste water system, treated as appropriate, and discharged to a point on the residual river network. Thus flow in the receiving water will, at that point, be artificially increased, chemically altered, and diurnally varied (Figure 2.4). Note that if the remote source were upstream of the urban area, natural river flows through the town would be reduced, but if the source were in another catchment, flows downstream of the town would be increased. The impact of water supply abstractions and waste water discharges is clearly dependent on the particular circumstances of the town. For example, a coastal town may have fewer waste water concerns, but bathing beaches to protect. Groundwater abstraction might also cause saline intrusion or subsidence.

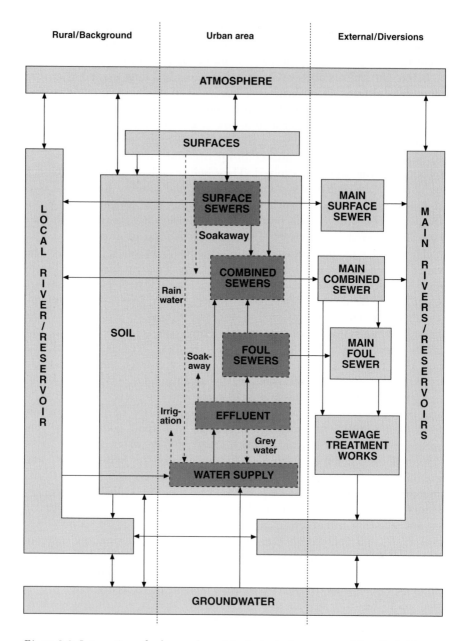

Figure 2.3 Interaction of urban and rural hydrological processes. (*Note*: 'leaky' boxes in SOIL box; dashed pathways in SOIL box mark water savings (and irrigation).

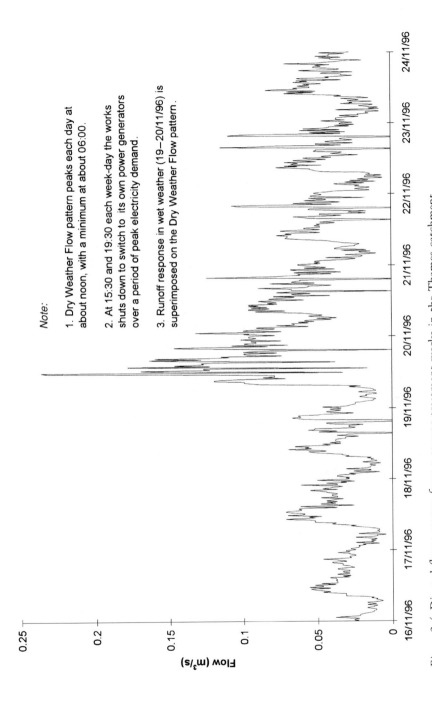

Figure 2.4 Diurnal flow patterns from a sewage treatment works in the Thames catchment.

Note:

1. Dry Weather Flow pattern peaks each day at about noon, with a minimum at about 06:00.

2. At 15:30 and 19:30 each week-day the works shuts down to switch to its own power generators over a period of peak electricity demand.

3. Runoff response in wet weather (19–20/11/96) is superimposed on the Dry Weather Flow pattern.

Flow (m³/s)

Consider now the urban drainage system. Increased paved and roof area will increase local runoff and require improvements to convey floodwater rapidly to the residual urban river network, itself extensively culverted. Rapid convergence through the drainage system of this increased runoff brings greater risk of flash flooding, particularly when downstream drainage capacity is (1) restricted by culvert dimensions, and (2) partially taken by groundwater infiltration. Moreover, particularly in older city centres, surface runoff and waste water drainage are usually provided by the one 'combined sewer' taking small runoff events to the sewage works, but with overflows provided so that higher flows (together with untreated sewage) can switch suddenly to pass directly to the river network. With growing population, increased per capita water use, and infiltration into old sewers, many over-flows work almost daily due to diurnal waste water patterns. In contrast, areas of newer development will usually have separate sewers, with just the surface runoff (misconnections excepted) permanently drained directly to local rivers. Urban runoff rates will often be four or more times greater than original rural rates, and heavily contaminated with a cocktail of surface washoff contaminants. The high imperviousness of urban areas means that the highest urban runoff rates occur with intense summer rainfall, when the available 'dilution' in natural watercourses is low due to the rainfall being retained in the dried soil profiles. Flow hydrographs in receiving water-courses will often include urban pulses superimposed on a longer lower swell (Figure 2.5). As these impacts become apparent, a range of engineering techniques will be used to reduce their effect on receiving waters. These techniques include underground storm tanks and oversize sewers with re-stricted outlets and 'hydrobrakes', flow diversions with real-time control or pumping, and surface storage ponds and washlands. Currently there is a growing interest in controlling runoff 'at source' by increasing local infiltra-tion, delaying runoff in porous pavements, gravel drains, or wetlands, and diverting surface water to non-potable domestic use. These methods reduce runoff rates and give time for settling and breakdown of pollutants. While the potential impact of urbanisation on flood runoff is large, with careful design it can often be eliminated or much reduced.

Finally, considering flood protection, main river channels through urban areas may be diverted, embanked, or provided with floodways. Such features may protect the urban area, but increase and speed up discharges to the down-stream environment. Similar effects may arise due to urban encroachment or bridge construction in the natural flood plain area. By contrast, a flood storage reservoir upstream of the urban area will protect the urban area and the downstream environment. The impacts are site dependent.

This discussion of urban impacts has considered water supply and runoff, but extensive urban developments will also affect climate and rainfall. The burning of fossil fuels and a reduction in evaporative cooling from vegetative

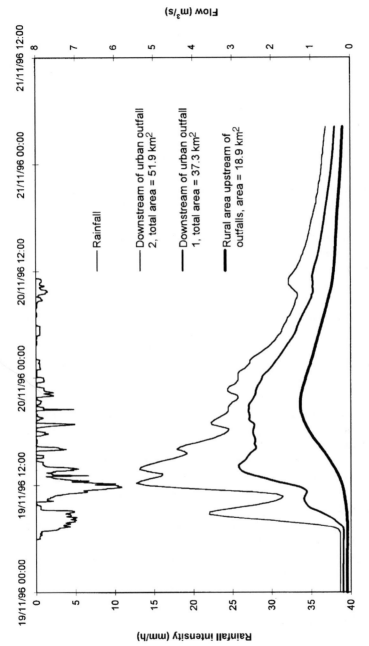

Figure 2.5 Flow upstream and downstream of urban outfalls in the Cut at Binfield.

surfaces may cause an 'urban heat island' which can result in increased convective rainfall.

More detailed consideration of urban impacts on catchment hydrology can be found in a range of reports, summaries, text books and design manuals, including Packman (1980), Hall (1984), Massing *et al.* (1990), Packman (1999), Chilton (1997).

2.8 Conclusions

The strategic issue for hydrologists is to assess the implication of land use change on water resources, flow regimes and water quality. At a time when such resources are coming under increasing stress, given recent droughts, and in the light of climate change scenarios indicating enhanced evaporation losses, it is particularly important to be aware of the potential dangers of encouraging land use change that may exacerbate water shortages. One example is the extension of irrigation, especially when applied as a spray. Another change that may reduce water resources yet further is extensive planting of forests with possibly much higher water use on major lowland aquifers. Less water also means less dilution of chemical pollutants.

Land use change over a region takes place slowly and in a largely piecemeal fashion as individual farmers respond to changes in legislation and market forces. The result is that as progressively larger areas are considered land use change impacts become attenuated and diluted, and in the case of groundwater there may be response-time delays of the order of tens of years. With regard to commercial forestry, the cropping cycle of British upland plantations is 40–60 years, so long-term and major cyclic variations are apparent in the impacts upon catchment hydrology, water quality and river sediment dynamics. Therefore management strategies can often be usefully focused upon the major opportunities to modify land use practices at the beginning and end of the forest rotation. Consequently, it is notoriously difficult to detect land use change impacts on streamflow or groundwater recharge for *large mixed* cover catchments especially with only short periods of record. This is despite the fact that *small-scale* process studies and experimental catchments may provide unequivocal evidence for a direct effect. There is the further difficulty in fully quantifying land cover changes when considering an extensive area owned by many individuals, and, even if a time trend in a hydrological parameter can be identified, its link to land use is often complicated by climatic variability. Developments in remote sensing appear to have great potential in this respect.

Although the vegetation cover of the land is controlled by its usage, it is the population, social, economic and political factors which control that usage and how intensive it is, and hence its impacts on hydrology. A better understanding of land use impacts on water resources will help policy makers and the water industry to judge the impacts of changes on their ability to meet the demands of domestic, industrial and agricultural consumers. Policy makers proposing significant change to the British countryside, through the CAP or woodland incentive schemes should be aware that there may be hydrological implications including low flows and groundwater. This consideration should not just be confined to national assessments but should be fed down through to local environment plans.

References

Barr, C.J. *et al.* (1993) *Countryside Survey 1990 Main Report. Countryside Survey 1990 Series, Vol. 2.* Department of the Environment, London.

Bell, M. (1992) The prehistory of erosion, in Bell, M. and Boardman, J. (eds), *Past and Present Soil Erosion.* Oxbow, Oxford. pp. 21–35.

Boardman, J. (1988) Severe erosion on agricultural land in East Sussex, UK, October 1987. *Soil Technol.*, 1, 333–48.

Boardman, J. (1995) Damage to property by runoff from agricultural land, South Downs, southern England, 1976–93. *Geog. J.*, 161(2), 177–91.

Boardman, J. (1998a) Modelling soil erosion in real landscapes: a western European perspective, in Boardman, J. and Favis-Mortlock, D. (eds), *Modelling Soil Erosion by Water.* Springer-Verlag, Berlin. pp. 17–29.

Boardman, J. (1998b) An average soil erosion rate for Europe: myth or reality? *J. Soil Water Conserv.*, 53(1), 46–50.

Boardman, J., Evans, R., Favis-Mortlock, D.T. and Harris, T.M. (1990) Climate change and soil erosion on agricultural land in England and Wales. *Land Degrad. Rehab.*, 2(2), 95–106.

Burt, T.P., Donohoe, M.A. and Vann, A.R. (1984) A comparison of suspended sediment yields from two small upland catchments following open ditching for forestry drainage. *Z. Fur Geomorph.*, N.S., 51–62.

Butcher, D.P., Labadz, J.C., Potter, W.R. and White, P. (1993) Reservoir sedimentation rates in the Southern Pennine region, UK, in McManus, J. and Duck, R.W. (eds), *Geomorphology and Sedimentology of Lakes and Reservoirs.* Wiley, Chichester. pp. 73–92.

Calder, I.R. (1992) Hydrologic effects of land use change, in Maidment D.R. (ed.), *Handbook of Hydrology*, Chap. 13. McGraw-Hill, New York.

Chilton, J. (ed.) (1997) Groundwater in the urban environment. Vol. 1: Problems, Processes and Management, in *Proceedings of the XXVII International Association of Hydrogeologists.* Congress on Groundwater in the Urban Environment.

DETR (1998a) Planning for the communities of the Future, Department of the Environment, Transport, and the Regions. Web page www.planning.detr.gov.uk/future/pcf003.htm.

DETR (1998b) *Land use and land cover.* www.environment.detr.gov.uk/epsim/ems8000.htm.

DoE (1992) *The UK Environment.* Government Statistical Service. Department of the Environment, HMSO, London.

DoE (1993) *Countryside Survey 1990 Summary Report.* Department of the Environment, HMSO, London.

DoE (1995) *The Occurrence and Significance of Erosion, Deposition and Flooding in Great Britain*. Department of the Environment, HMSO, London.

DoE (1996) *Digest of Environmental Statistics No. 18*. Department of the Environment, HMSO, London.

Duck, R.W. (1985). The effect of road construction on sediment deposition in Loch Earn, Scotland. *Earth Surface Processes Landforms*, 10, 401–406.

Evans, R. (1990a) Soil erosion: its impact on the English and Welsh landscape since woodland clearance, in Boardman, J., Foster, I.D.L. and Dearing, J.A. (eds), *Soil Erosion on Agricultural Land*. Wiley, Chichester. pp. 231–254.

Evans, R. (1990b). Soils at risk of accelerated erosion in England and Wales. *Soil Use Manag.*, 6(3), 125–31.

Evans, R. (1992) Erosion in England and Wales – the present the key to the past, in Bell, M. and Boardman, J. (eds), *Past and Present Soil Erosion*. Oxbow, Oxford. pp. 53–66.

Evans, R. (1993) Extent, frequency and rates of rilling of arable land in localities in England and Wales, in Wicherek, S. (ed.), *Farm Land Erosion: In Temperate Plains Environment and Hills*. Elsevier, Amsterdam. pp. 177–190.

Evans, R. (1996) *Soil Erosion and Its Impacts in England and Wales*. Friends of the Earth, London.

Evans, R. (1997) Soil erosion in the UK initiated by grazing animals: a need for a national survey. *Appl. Geog.*, 17(2), 127–41.

Favis-Mortlock, D., Boardman, J. and Bell, M. (1997) Modelling long-term anthropogenic erosion of a loess cover: South Downs, UK. *The Holocene*, 7(1), 79–89.

Forestry Commission/Countryside Commission (1996) *Woodland Creation: Needs and Opportunities in the English Countryside*. Countryside Commission, Northampton.

Forestry Commission (1993) *Forests and Water Guidelines*, 3rd edn. HMSO, London.

Foster, I.D.L., Grew, R. and Dearing, J.A. (1990) Magnitude and frequency of sediment transport in agricultural catchments: a paired lake-catchment study in midland England, in Boardman, J., Foster, I.D.L. and Dearing, J.A. (eds), *Soil Erosion on Agricultural Land*. Wiley, Chichester. pp. 153–71.

Foundation for Water Research (1994) *Urban Pollution Management Manual*. Foundation for Water Research, Marlow, Buckinghamshire.

Green, F.H.W. (1980) Field Underdrainage before and after 1940. *Agric. History Rev.*, 28, 120–123.

The *Guardian* (1997) Most of the World's Forests are Gone. The *Guardian*, 9 October 1997, p. 7.

Hall, M.J. (1984) *Urban Hydrology*, Elsevier, Amsterdam.

Hall, R.J. and Allen, S.J. (1997) Water use of poplar clones grown as short-rotation coppice at two sites in the UK. *Aspects Appl. Biol.*, 49, 163–72.

Harding, R.J. *et al.* (1992) Hydrological impacts of broadleaf woodlands: implications for water use and water quality. *National Rivers Authority Report 115/03/ST*. HSMO, London.

Heather, A.I.J. and Carter, A.D. (1996) Herbicide losses from hard surfaces and the effect on ground and surface water quality, *Aspects Appl. Biol.*, 44, 1–8.

Heathwaite, A.L., Burt, T.P. and Trudgill, S.T. (1993) Overview – the nitrate issue, in *Nitrate: Processes, Patterns and Management*, (ed.) T.P. Burt, A.L. Heathwaite and S.T. Trudgill, Wiley, Chichester.

HMSO (1995) *Rural England – A nation Committed to a Living Countryside*. CM 3016. HMSO, London.

House of Commons (1996) *Environment Committee First Report. Water Conservation and Supply*. HMSO, London.

Johnson, R.C. (1991) Effects of upland afforestation on water resources: The Balquhidder Experiment 1981–1991, *Institute of Hydrology Report No. 116*. Institute of Hydrology, Wallingford.

Kirby, C., Newson, M.D. and Gilman, K. (1991) Plynlimon research: the first two decades. *Institute of Hydrology Report No. 109*. Institute of Hydrology, Wallingford.

Leeks, G.J.L. (1992) Impact of plantation forestry on sediment transport processes, in Billi, P., Hey, R.D., Thorne, C.R. and Tacconi, P. (eds), *Dynamics of Gravel Bed Rivers*. Wiley, Chichester, pp. 751–70.

Leeks, G.J.L. and Marks, S.D. (1997) Dynamics of river sediments in forested headwater streams: Plynlimon, Mid Wales. *Hydrol. Earth Syst. Sci.*, 1(3), 483–97.

MAFF (1996) *British Survey of Fertiliser Practice: Fertiliser Use on Farm Crops for Crop Year 1996*. MAFF Publications, London, p. 69.

MAFF (1997) *Pesticide Usage Survey Report 100: Review of Usage of Pesticides in Agriculture and Horticulture Throughout Great Britain 1984–1994*. MAFF Publications, London, p. 19.

MAFF (various dates) *Agricultural Returns of Great Britain or Agricultural Statistics, UK*. HMSO, London.

Maitland, P.S., Newson, M.D. and Best, G.A. (1990) *The Impact of Afforestation and Forestry Practice on Freshwater Habitats*. Nature Conservancy Council, Focus on Nature Conservation, No. 23, Peterborough.

Marks, S.D. and Rutt, G.P. (1997) Fluvial sediment inputs to upland gravel bed rivers draining forested catchments: Potential ecological impacts. *Hydrol. Earth Syst. Sci.*, 1(3), 499–508.

Massing, H., Packman, J.C. and Zuidama, F. (eds) (1990) *Hydrological Processes and Water Management in Urban Areas*. IAHS Publication, 198, Wallingford.

Neal, C. (ed.) (1997) Water quality of the Plynlimon catchments (UK) Special Issue. *Hydrol. Earth Syst. Sci.*, 1(3), 381–764.

Newson, M.D. (1980) The erosion of drainage ditches and its effects on bedload yields in Mid-Wales, *Earth Surf. Proc. Landforms*, 5, 275–90.

Newson, M.D. (1989) Flood effectiveness in river basins: progress in Britain in a decade of drought, in Beven, K. and Carling, P. (eds), *Floods: Hydrological, Sedimentological and Geomorphological Implications*. Wiley, Chichester. pp. 151–69.

Newson, M.D. and Leeks, G.J.L. (1987) Transport processes at the catchment scale: a regional study of increasing sediment yield and its effects in mid-Wales, UK, in Thorne, C.R., Bathurst, J.C. and Hey, R.D. (eds), *Sediment Transport in Gravel Bed Rivers*. Wiley, Chichester. pp. 187–224.

NRA (1992) *The Influence of Agriculture on the Quality of Natural Waters in England and Wales*. Water Quality Series No. 6, National Rivers Authority, HMSO, London. p. 154.

NRA (1994) *Water: Nature's Precious Resource*. National Rivers Authority, HMSO, London.

NRA (1995) *Pesticides in the Aquatic Environment*. Water Quality Series No. 26, National Rivers Authority, HMSO, London, p. 92.

Packman, J.C. (1980) The effect of urbanisation on flood magnitude and frequency, *Institute of Hydrology Report No. 63*. Institute of Hydrology, Wallingford.

Packman, J.C. (1999) Flood estimation in mixed urban/rural catchments, Institute of Hydrology Report (in preparation).

Robinson, M. (1986) The extent of farm underdrainage in England and Wales, prior to 1939. *Agric. History Rev.*, **34**, 77–85.

Robinson, M. (1998) 30 years of forest hydrology changes at Coalburn: water balance and extreme flows. *Hydrol. Earth Syst. Sci.*, 2(2), in press.

Robinson, M. and Armstrong, A. (1988) The extent of agricultural field drainage in England and Wales, 1971–80. *Trans. Inst. British Geographers*, **13**, 19–28.

Robinson, M. and Rycroft, D.W. (1999) The impact of drainage on streamflow, in van Schilfgaarde, J. and Skagg, R.W. (eds) *Agricultural Drainage*. American Society of Agronomy, Madison, 753–86.

Slattery, M.C., Burt, T.P. and Boardman, J. (1994) Rill erosion along the thalweg of a hillslope hollow: a case study from the Cotswold Hills, central England. *Earth Surf. Proc. Landforms*, **19**, 377–85.

Soil Survey of England and Wales (1983) *Soil Map of England and Wales*. Soil Survey of England and Wales, Harpenden.

Soil Survey and Land Research Centre (1993) *Risk of Soil Erosion in England and Wales by Water on Land under Winter Cereal Cropping*. Soil Survey and Land Research Centre, Silsoe.

Stott, T. (1989) Upland afforestation, does it increase erosion and sedimentation. *Geog. Rev.*, March 1989, 30–32.

Stretton, C. (1984) Water supply and forestry, a conflict of interests: Cray Reservoir, a case study. *J. Inst. Water Eng. Sci.*, **38**, 323–30.

UN Centre for Human Settlements (1996) *An Urbanising World: Global Report on Human Settlements*, Oxford University Press, Oxford.

Walling, D.E. (1990) Linking the field to the river: sediment delivery from agricultural land, in Boardman, J., Foster, I.D.L. and Dearing, J.A. (eds), *Soil Erosion on Agricultural Land*. Wiley, Chichester. pp. 129–52.

Ward, N., Clark, J., Lowe, P. and Seymour, S. (1993) *Water Pollution from Agricultural Pesticides. Research Report*. Centre for Rural Economy, University of Newcastle upon Tyne, p. 80.

3

RIVER CHANNEL MODIFICATION
IN THE UK

David Sear, David Wilcock, Mark Robinson and
Karen Fisher

This chapter considers the history of direct channel modification in the UK and the administrative responsibilities of various organisations. It details the associated hydrological and geomorphological impacts and the attempts made to enhance and rehabilitate degraded channels.

3.1 Introduction

3.1.1 *Natural watercourses*

Totally natural watercourses differ substantially from river channels commonly seen today in the UK (Plate 3.1). Indeed, for many rivers, the very fact that we can observe the actual watercourse directly means that that they are not natural, since most of UK was originally wooded. This difference between natural and today's modified water courses can be characterised in four ways: changes in channel form; changes in rates of process; the role of vegetation; and connectivity with floodplains. Each of these is inter-related and contributes to modifying the hydrological response of the catchment by changing channel capacity and flow resistance. Natural channels are complex, dynamic systems. They are free to adjust their form and flow conveyance at rates, in directions and at locations determined by the natural properties of its bed and banks, including riparian vegetation. This adjustment takes place over a range of time scales. There is also strong interaction between river channels and their adjacent floodplains. This contrasts markedly with modified river systems where form is simplified, diversity is reduced, adjustment and dynamism is controlled, vegetation is managed or removed and floodplains are hydrologically disconnected from the channel.

Hydrological processes in natural catchments are dominated by climate, lithology, relief, soil type and land cover. This leads to a wide range of river types. At one extreme are rivers whose flow responds rapidly to rainfall. These

Plate 3.1 Comparative photographs of a natural (a) and modified (b) UK river channel illustrating the differences in vegetation and morphological diversity (David Sear).

are often found in upland catchments, underlain by impermeable rocks such as schist, or in the lowlands having impermeable soils, such as clay. At the other extreme are intermittent or perennial groundwater-dominated rivers developed on permeable lithologies, such as chalk (Sear *et al.*, 1999). Over long time periods, natural landcover may change and modify catchment hydrology. In addition, natural perturbations in climate produce major shifts in catchment hydrology and river form over decadal and longer timescales (Arnell, 1996).

Climate and the physical character of the catchment influence natural river channel processes in the following way. The slope and discharge determine available energy and the bed sediments, bank materials and vegetation determine channel boundary properties. In turn, slope, sediments and vegetation are often influenced by past events such as glaciations and high-magnitude floods. The form of a river therefore results from the interaction of these different influences. As channel form alters, in response to changes in discharge regime or delivery of sediments to the river network, there can be substantial changes in flow resistance. This is particularly the case where channels run through wood areas, where a fallen tree may block the channel completely. The great spatial variability in channel form and process creates diverse river habitats for wildlife within the channel and on the floodplain, characterised by heterogeneity or patchiness. Consequently, unmodified rivers tend to be morphologically and hydrologically unique; this is one of the main criteria used to define the conservation value of natural systems (Gregory, 1997). In rivers with hydrologically connected floodplains, the patchwork of habitats forms a complex and diverse ecosystem with many self-regulating functions that have benefits for flood peak attenuation, sediment storage and nutrient recycling. It is the removal of these functions by channel modification that has the greatest impact in UK river systems and that creates the greatest challenge for river restoration.

3.1.2 Channel modification

Channel modifications refer to those management activities that alter the form of the river channel, specifically affecting the planform, cross-section and long-profile. Figure 3.1 illustrates the nature of the most common forms of modification in the UK. In general, channel form is modified as part of extensive land drainage schemes that required improvement in the efficiency of the river network, and as flood protection to confine high flows within the river network (Brookes, 1988; Robinson, 1990; Sear *et al.*, 1995). Such modifications have traditionally been grant-aided by the Ministry of Agriculture, Fisheries and Food (MAFF), Scottish Office or the Department of Agriculture for Northern Ireland (DANI) after being judged to be technically sound and of economic benefit. Associated with these modifications is a general requirement to maintain the design conveyance or standard of

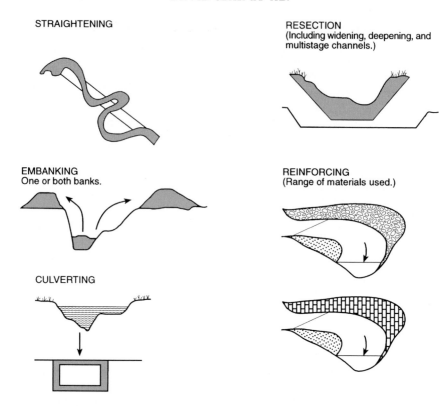

Figure 3.1 Common forms of direct channel modification undertaken for flood protection and land drainage. Other maintenance modifications might include desilting, shoal removal, bank repair and heavy vegetation removal (including roots and sediment).

flood protection. Maintenance represents a continual disturbance of the river channel, often annually, and may significantly modify the channel form. This often involves the removal of sediment accumulations or the reinforcement of river channel banks and beds where erosion threatens land or specific structures, for example a river embankment or line of communication. This has been termed sediment-related river maintenance by Sear *et al.* (1995) and is distinct from weed clearance and debris removal that are also part of this maintenance process. An exception to these practices are the channel modifications arising from the winning of gravels directly from the river bed for building aggregate (Sear and Archer, 1998) and more recently, direct channel modifications undertaken in order to enhance instream habitats and to increase morphological diversity through river rehabilitation and restoration (Brookes and Shields; 1996, Sear *et al.*, 1998).

3.2   The history of UK river channel modification

Channel modification within the UK has distinct regional and geographical trends, driven partly by the nature of the land and rivers themselves as well as by the history of water management legislation. Among the earliest forms of channel modifications in the UK were those undertaken by the Romans who during their occupation constructed drainage dikes and embankments, notably along the margins of the Fens. The extent of drainage and flood protection works increased during the medieval period to such an extent that specific legislation was required to ensure maintenance of water resources for milling and fisheries through the establishment of 'Courts of Sewers'. The main era of channel modification began in the seventeenth century with the introduction of technological expertise from Holland coupled with private investment from wealthy land owners. At first, low-lying lands subject to inundation were the focus of major drainage schemes featuring the construction of new watercourses and extension of the river network, together with dredging of existing watercourses to improve drainage. In the uplands, watercourses were often diverted or realigned for milling purposes. In the nineteenth century, railway construction began to impact on river channels. Extensive gravel extraction occurred in many UK rivers in order to provide material for construction of railway embankments and associated infrastructure, potentially resulting in incision and planform instability (Sear and Archer, 1998). At the same time, where rivers impinged on the line of the railway, channel reinforcement and planform modification occurred, leading to expensive and prolonged maintenance commitments. During the Highland Clearances, Napoleonic Wars and the two World Wars, prisoners of war were used to clear large sections of watercourses, reinforce banks and remove riparian vegetation in order to stabilise channels and improve drainage. Channel modification on some rivers in UK sometimes only involved minor bush clearance and shoal removal work. Although generally discouraged, shoal removal by private land owners or tenants is widespread, particularly in upland gravel-bed rivers where materials may be used for track repairs. The cumulative impacts of such small-scale activity are not known, although larger-scale works can lead to channel instability and downstream sediment starvation (Newson and Leeks, 1987; Sear *et al.*, 1995). Until recently, many large estates had workforces that routinely spent the autumn maintaining river banks and embankments, the impacts of the decline in this activity being similarly unknown. Arguably the period of most intensive and extensive channel modification occurred across the UK between 1930 and 1990, driven initially by war-time demand for increased agricultural output. This was sustained by the CAP and funding for land drainage improvements to ensure high rates of agricultural productivity. Early channelization schemes were typically designed to accommodate the five-year flood in agricultural areas. Since the level of protection is related to economic costs and benefits,

MAFF provided guidance on the levels of service associated with different types of agricultural land. These varied from one in one (1:1) year protection for extensive grazing land to one in ten (1:10) years where high-value root crops, cereals or grasses were grown. Where channelised rivers passed through urban areas or close to industrial premises, however, flood protection was sometimes provided against 1:200 year floods (Hutton, 1972). Even in small towns, channel modification works in the UK are designed to protect against the 1:100 year flood (Northern Ireland Department of Agriculture, 1992) whilst properties may benefit from protection against 1:30 to 1:50 year flows. Thus for agricultural areas, design protection may vary from 1:1 to 1:10 years and may therefore have little effect on very high return interval floods and may require relatively limited channel modification. In comparison, channels in urban areas designed to convey high return interval floods are often highly modified and require reinforcement. A complication to this picture is that in the past many channels were modified to carry a particular discharge rather than a particular return interval. In addition, maintenance activity is seldom based on exact estimation of real need, and thus channels can be over or under maintained. Attempts at standardising river maintenance have been made founded on the notion of a standard of service using the value of the land and the nature of the land use (Fitzsimmons and Pimperton, 1990).

Since the early 1980s, the recognition that channel modifications had substantially impacted the natural river environment of the UK brought about a change in emphasis towards protecting and (later) enhancing the river environment. In the late 1990s, river rehabilitation and enhancement are rapidly becoming the dominant form of channel modification process. In an ideal situation, the goal of river rehabilitation is to return the river to its pre-disturbance state. Defining that state is difficult for two reasons, first the length of time during which catchment and channel modifications have been made makes definition of pre-disturbance state difficult and secondly, the lack of information on natural river channels makes the rehabilitation of hydrological and hydraulic regime problematic. In many instances, rehabilitation may be achieved through natural recovery where the river still retains sufficient energy to modify its boundary. However, in low energy, cohesive channels, some form of active restoration will be required to recreate appropriate features. In both cases, it is vital to include the floodplain within the project since it is from the connectivity of channel and floodplain that many of the greatest benefits of rehabilitation are derived. These benefits are broad ranging, affecting catchment management as a whole (Table 3.1). Recognition of the wider potential benefits will stem from improved understanding of the contribution that rehabilitation can make towards better management of floods, droughts and water quality and the achievement of governmental strategies for improved biodiversity, CAP reforms and agri-environmental support schemes.

Table 3.1 Benefits arising from river restoration.

Function	Benefit
Ability to adjust to changes in sediment supply and discharge regime	Reduced maintenance, ability to respond to external changes (e.g. climatic or land/channel management)
Increased habitat diversity	Species richness and biodiversity
Flood water storage on floodplains	Flood protection/benefits for wading birds
Attenuation of floodwave	Flood protection/reduced downstream erosion
Physical diversity during low-flows	Maintenance of water depth/water quality during low flows
Increased residence time of sediments	Reduced maintenance in downstream reaches
Increased nutrient cycling in floodplains/channels	Improved water quality
Increase in aesthetic value of watercourse	Increased quality of life

3.3   Administration of channel modification

In England and Wales, the Environment Agency has permissive powers to undertake land drainage and flood protection works on designated 'main rivers'. Funding for such activity comes from direct grant-aid from MAFF as well as from the levying of precepts on local councils and internal drainage boards, the latter source largely paying for river maintenance. In addition to the Environment Agency, channel modifications have been and continue to be undertaken by the internal drainage boards that cover approximately 8% of England and Wales, maintaining some 27 000 km of intermediate watercourses that usually discharge to the main river network (Robinson, 1990). Local authorities, both county and district, also have powers to carry out channel modifications on non-main rivers, although in comparison these are few, and often involve consultation with the Environment Agency. Riparian owners have some responsibility for erosion control and maintenance of both main and non-main watercourses and have maintenance responsibility for the enormous network of minor watercourses and ditches that often make up the headwater areas of river catchments. In addition to these bodies, many non-governmental organisations (NGOs) such as the Royal Society for the Protection of Birds and County Wildlife Trusts have interests in channel modifications and the preservation and enhancement of natural channel habitat. For England and Wales, the management of channel modifications on main river has changed radically in the last decade from single-function, single-funded channel works to multi-functional, environmentally aligned works designed to provide the 'Best Practicable Environmental Option' (BPEO) and charged with incorporating sustainability objectives within their goals. However, the degree to which this management philosophy extends to other

Table 3.2 Statutory opportunities for channel modification in Scotland.

Private Rights whereby riparian owners can directly modify the river channel provided the quantity and quality of water upstream or downstream of their land is not impacted.

War-time Land Drainage Acts (legislation to be repealed in April 1999) that control the maintenance of 13 schemes conducted in the 1940s for agricultural improvement at an annual cost of £120 thousand pounds per annum.

Land Drainage of Scotland Acts (1958) that permit one or more farmers to produce a drainage proposal to the Scottish Office to enhance agricultural activity.

Flood Prevention of Scotland Act 1961 gives Local Authorities powers to modify the channel to prevent flooding of communities. This may include works on river channels remote from communities where this can be shown to benefit the flood protection of the community itself. Subject to planning consent and design consent from Scottish office.

Fisheries Interests in particular the District Salmon Fisheries boards that have statutory rights to modify the channel for the benefit of salmonid fisheries. In practice, this usually means enhancement features and gravel redistribution.

---

organisations remains variable (Newbold *et al.*, 1983). The current trend is towards the enhancement of channel habitat and the environmentally sensitive maintenance of existing modified watercourses. Legislation supporting such activities is available within the Environment Act (1996) and within evolving EU directives. New land drainage is now relatively rare, and most modification is now focused on urban flood protection and the maintenance of existing drainage and flood alleviation schemes. Rehabilitation of degraded watercourses which began in the early 1980s (Brookes, 1988) has, by contrast, become one of the major causes of channel modification in England and Wales.

The history of channel modifications in Scotland is not well documented, in part the result of a fragmented water management system but also arising from the interplay of powerful lobbies, notably the salmon fisheries concerns and agriculture. Channel modifications occurred as early as the fourteenth century in the Borders, associated with the drainage of floodplains and the extraction of marl (Gilvear, D.J., personal communication). At present a number of opportunities for channel modification exist in Scotland (Table 3.2) which have led to relatively few large schemes (affecting <2% rivers) but widespread smaller schemes designed to improve local farmland or enhance fisheries interests. Changes in the approach to channel modification are planned including the repeal of the war-time land drainage acts, whilst large scale restoration initiatives are evolving through NGOs such as World Wildlife Fund (WWF). Responsibility for channel modifications currently resides with The Scottish Office, riparian owners, local authorities, salmon fisheries boards and the Scottish Environment Protection Agency (SEPA).

In Northern Ireland, the combination of relatively high annual rainfall (1075 mm), low evapotranspiration (430 mm) and extensive areas of impermeable soils (Cruickshank, 1997) has made flooding and poor drainage a recurrent theme of the country's social and economic history for at least the last 150 years (Dooge, 1987). Systematically designed and engineered channel modification, designed to relieve flooding and/or to provide increased channel capacity for discharges from extensive field drainage schemes, goes back only 50 years and usually proceeded upstream in any particular catchment. In addition to land drainage and flood protection works, strong fishing interests maintain extensive 'enhancements' in many salmon rivers, often involving redistribution of gravel shoals, channel narrowing to scour silts and the construction of instream features designed to create scour pools and deflectors. Responsibility for river channel modifications currently resides with DANI, guided by legislation set out in the Drainage (NI) Order 1973. Under this legislation, a Northern Ireland Drainage Council decides which watercourses should be designated for drainage, the Council's membership being drawn from a wide variety of interests. The DANI retains authority for proposing and implementing individual drainage schemes operationalised by the Rivers Agency, a 'next steps' agency within the Department.

3.4   The UK extent of channel modifications

3.4.1   General

The nature and extent of direct channel modification in the UK reflects the broad physical geography of the country with distinctive variations between upland and lowland watercourses (Table 3.3). The influence of climate and land use within these areas are again important factors determining the geography of modifications and the nature of their maintenance (Newson and Sear, 1994). Determining the geography of channel modifications has been made possible by the recent publication of the UK Environment Agency River Habitat Survey (RHS). The RHS is a systematic framework for the collection and analysis of data associated with the physical structure of watercourses. Data collection is based on standard 0.5 km reaches sampled on a random stratified basis across the UK, so that each 10 km grid square has within it three RHS sites. Operator and seasonal bias have been accounted for in the data collection. The national picture emerging from an analysis of the RHS data (Environment Agency, 1998) and shown in Figure 3.2, reveals a widespread distribution of rivers that have experienced one or more of six main channel modifications: straightening, resectioning, reinforcing, embanking, culverting and the construction of weirs and sluices. As Table 3.3 and Figure 3.2 reveal, the majority of modified channels are associated with lowland regions of the UK, where agricultural drainage, communications networks and urban centres all contribute to a long history of river channel management. In the uplands, reinforcement of channel boundaries represents the dominant form of modification, reflecting the high energy and dynamic nature of rivers in this category. The data for straightened channels is known

63

Table 3.3 Proportion of UK river channels with evidence of modification based on UK Environment Agency River Habitat Survey (Environment Agency, 1998).

Modification	UK upland	UK lowland	England and Wales upland	England and Wales lowland	Scotland upland	Scotland lowland	Northern Ireland lowland
Straightened	0.0	6.2	0.0	7.3	0.0	0.5	0.4
Resectioned	12.5	44.0	10.1	45.6	17	32.1	53.5
Reinforced	35.4	51.9	41.6	52.0	23.1	44.8	66.3
Embanked	5.1	14.9	5.2	14.3	4.4	12.1	31.0
Culverted	3.5	9.3	5.2	10.5	0.0	4.8	3.1
Weirs or sluices	8.4	15.0	10.6	15.6	0.0	4.8	3.1
Number of reaches surveyed	593	5091	404	4155	181	588	262
Total survey distance (km)	297	2546	202	2078	91	294	131

Data derived from the UK Environment Agency River Habitat Survey, 1998. Note that low survey density in Scotland and NI may affect overall estimates.

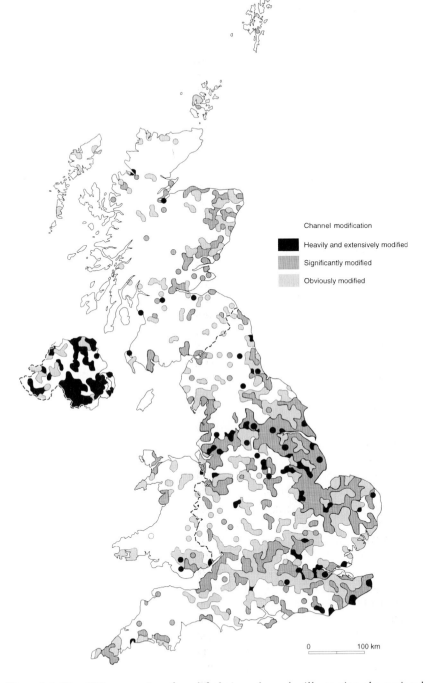

Channel modification

■ Heavily and extensively modified

▨ Significantly modified

░ Obviously modified

0   100 km

Figure 3.2 The UK geography of modified river channels, illustrating the regional variations associated with lowland and urban river management. (Data from UK Environment Agency River Habitat Survey.)

Capital works and major
improvement schemes

Main rivers—mostly maintained

0   100 km

(a)

Figure 3.3 (a) The extent of capital works and maintained watercourses undertaken
between 1930 and 1980. (Figure from Brookes *et al.*, 1983.) (b) The
geography of river maintenance in England and Wales (Sear *et al.*, 1995).
(c) River rehabilitation schemes in the UK.

to be inaccurate, since many reaches in upland and lowland UK have been
straightened, although as this dataset suggests, this is not as extensive as
other forms of modification. There are clear regional variations; in particular,
Northern Ireland is shown to have the highest proportions of modified channels

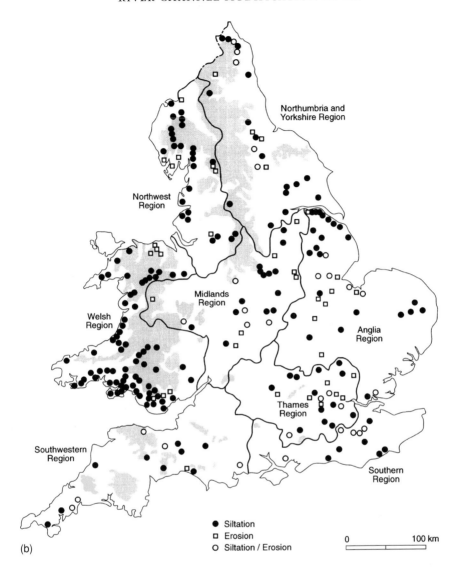

Northumbria and
Yorkshire Region

Northwest
Region

Welsh
Region

Midlands
Region

Anglia
Region

Thames
Region

Southwestern
Region

Southern
Region

● Siltation
□ Erosion
○ Siltation / Erosion

0   100 km

(b)

associated with agricultural land drainage practices, whilst the relatively
densely populated regions of England account for the higher proportions of
culverted channels. The cause of these modifications is in part explained by
Figure 3.3(a) which illustrates the extent of maintenance and capital works
within England and Wales for the period 1930–80 as recorded in the minutes of water management land drainage and flood defence reports (Brookes
et al., 1983). This amounts to over 35 500 km of main river, of which some
4500 km have reinforced banks. The continued legacy of these modifications

67

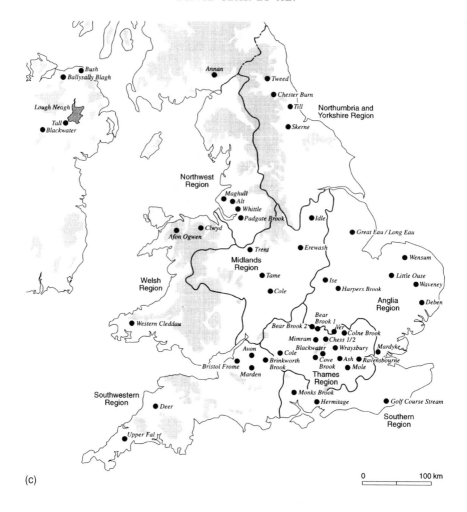

(c)

0   100 km

is shown in Figure 3.3(b) in the form of sediment-related river channel maintenance (Sear *et al.*, 1995). The number of sites is only a sample, estimated at around 15% of the total maintenance commitment of the Environment Agency. No comparative information was available for Scotland and Northern Ireland. In Northern Ireland since 1947, arterial drainage (channelisation) of more than 6 000 km of river channel across the Province has occurred, at a regional density of 0.34 km km^{-2}, higher than most of the regional densities quoted by Brookes (1988) and supported by the values given in Table 3.3. The regional disparity in channel modifications depicted in Figure 3.2 reflects the focus of land drainage in the lowlands of eastern Britain and the protection of urban centres. In England and Wales, this has resulted in significant regional variations (Table 3.4) with the Anglian region returning the highest

Table 3.4 UK Environment Agency regional variations in channel modifications within England and Wales.

	All	Northwest	Northeast	Midland	Anglian	Welsh	Thames	Southwest	Southern
Embanked	42.6	41.3	41.9	42.7	50.4	38.5	41.3	39.0	43.2
Resectioned	40.2	36.3	41.7	43.6	66.0	19.1	65.4	30.8	49.7
Culverted	10.1	10.7	11.2	10.2	9.7	5.1	13	8.4	20.2
Reinforced	50.9	63.5	43.9	51.2	37.3	49.1	56.5	53.4	46.0

Emboldened figures represent largest value. Data derived from the UK Environment Agency River Habitat Survey, 1998.

proportions of embanked and resectioned river channels. Regions with significant upland areas return the highest proportion of reinforced channels that reflect attempts to constrain lateral channel migration in dynamic gravel-bed rivers, an exception being the Thames region where the overriding influence is the presence of a major urban centre. In contrast, the low-energy river channels in the Anglian region do not require substantial reinforcement despite resectioning and embanking. The Welsh region, with the lowest proportion of modified channels, has a river network of high habitat quality. It is important to note that the extent of modification to the UK river network explored above does not include those adjacent reaches which have experienced adjustment, or that are impacted hydrologically by the presence of structures. Thus the true extent of modified river channels will be much greater. To summarise, the national picture of channel modification is one of long-term intervention extending over much of the river network, with continual (though reducing) levels of maintenance, concentrated in the impermeable lowlands and in major urban areas.

3.4.2 *River rehabilitation and enhancement*

By way of contrast to the picture of UK channel modifications arising from development of the floodplain, Figure 3.3(c) illustrates the distribution of river rehabilitation and enhancement sites in England and Wales that amount to less than 50 km (<0.01%) of main river channel. The extent of rehabilitation captured in Figure 3.3(c) reflects the early uptake and slow transmission of practice across the water management sector and within the Environment Agency specifically. The majority of river rehabilitation schemes have been undertaken by the UK Environment Agency using flood defence maintenance and capital works budgets with 'windfall' funding accounting for many of the other NGO schemes. This partnership model is likely to dominate the future of river rehabilitation in the UK, resulting in wider public opportunity to become involved, although the control on the appropriateness of channel modifications are less certain, with many fishing clubs attempting rehabilitation for single function objectives.

The Institute of Hydrology has adapted the Physical Habitat Simulation (PHABSIM) system, originally developed in USA, to assess the success of channel restoration. PHABSIM was employed on the restored reach of the River Wey in Surrey (Acreman *et al.*, 1996). It showed that although channel modifications had increased the diversity of depth and stream flow velocity, these were insufficient to improve the physical habitat available for trout and chub. The study demonstrated the need for a technique that could estimate physical habitat availability at the design stage. Dunbar *et al.* (1997) linked standard hydraulic models, used for channel design (ISIS, MIKE 11, HEC-RAS) to the hydro-ecological models in PHABSIM to produce a predictive tool to aid channel rehabilitation.

3.5   Geomorphological impacts of UK river channel modification

The impacts of channel modifications necessarily depend on the type of modification, thus embanking a river channel may not alter the function of the channel processes, but influences the total discharge range contained within the reach. More clearly, there are different impacts between those modifications associated with improving channel conveyance for drainage and flood protection and those conducted for enhancement or restoration. Scale effects are also evident, particularly in integrated land and arterial drainage schemes. Different processes may be important as different sizes of catchment are considered, e.g. the increase in storage resulting from underdrainage of floodplains may locally decrease peak flows from a small catchment, but when linked to efficient arterial drainage channels in a larger catchment, may lead to increased flood peaks downstream owing to the faster travel time of flood waters from the upper catchment (Newson and Robinson, 1983; Robinson, 1990; Robinson and Ryecroft, 1998). Thus, when considering the impacts of channel modifications, it is important to have regard for the location of the reach, the extent and the type of modification.

Morphological changes resulting from channel modification depend on the reason for the modification and the nature of the river channel, e.g. straightening meanders may be an option for improving conveyance, but the nature of the drainage requirements may actually necessitate lowering of the river bed. The options available will also vary depending on the geotechnical properties of the river sediments. In major channelisation schemes, changes in channel morphology have often been dramatic. On the River Main drainage scheme in County Antrim, for example, the channel bed was lowered by three metres in places and much of the main river channel straightened and resectioned. Individual reaches were, sometimes, entirely relocated in the flood plain and several meandering reaches eliminated altogether. Over the 5.25 km reach illustrated in Figure 3.4, 43 pools were eliminated and the river length shortened by 23%. These changes were all in a direction which increased slope and hydraulic radius, and reduced channel roughness, thus bringing about the increased flow velocities required to evacuate water from the land more efficiently than before. Such channel modifications, especially those which confine flood flows of a five, ten, fifty year return period within channels formerly adjusted to smaller bankfull floods, increased stream power to such an extent that bed and bank protection works – sheet piling, rock armouring, etc. – were often engineered into these schemes at the design stage, thus destroying the adjustability of form in relation to streamflow and sediment transport rates inherent in natural river systems. In the absence of such reinforcement, river channels may adjust their form over time, although maintenance may prevent this through removal of vegetation and sediment accumulations. Sear *et al.* (1995) document the morphological

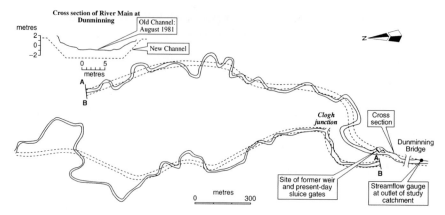

Figure 3.4 The planform of a 5.25 km stretch of the River Main, in Northern Ireland, before and after channelisation. The pre- and post-channelisation cross-section of the river, at the downstream end of this particular reach, is also shown. Note that the total length of river is presented as two separate reaches for ease of presentation. (After Wilcock and Essery, 1991.)

response of a low-energy clay lithology stream to major regrading. The River Sence, in this example, was not straightened, but had the river bed lowered and the banks pulled back to improve land drainage. Subsequently, major bank failures occurred throughout most of the improved watercourse arising from geotechnical failure of the bank materials in response to bed lowering and an increase in bank height and angle. The increased sediment supply from the exposed banks led to accumulation in the river bed and precipitated a programme of maintenance.

The indirect impacts of channel modifications have been outlined by Brookes and may account for extensive downstream and upstream channel adjustments. In a survey of 57 sites of channel modification, Brookes (1987) observed significant downstream adjustments characterised by channel enlargement associated with bank instability in high-energy gravel-bed rivers. In low-energy river systems, morphological adjustments were characterised by accumulations of sediment, reducing channel capacity in the modified reach, and raising bed elevations upstream. The extent of these impacts can be significant, with many between 1–5 km downstream of the modified reach. The response to such adjustment is often the reinforcement or maintenance of the reach, thus extending the length of modified channel still further. Upstream of modified reaches, increase in water surface slope caused by drawdown through the reach can initiate incision and bank failure. Alternatively, elevated flood levels resulting from embankments may induce backwatering and sediment accumulation.

Dredging or restoration of new channels greatly enhances sediment loads as a result of exposure of large areas of bare, often unconsolidated, substrate. The general environmental effects of enhanced sediment loads in the UK's aquatic environments were initially described by Swales (1982). These variously include: increased turbidity, reduced light penetration and a decline in photosynthetic productivity; damage to the feeding apparatus of filter feeders such as caddis flies and abrasion to the gill tissue in fish; transformation of cobble-bedded into silt-bedded streams; and increased nutrient loading due to the adsorption on suspended colloids of phosphates and ammonia. During channelisation on the River Main, suspended sediment concentrations appeared to increase between five and ten fold for all discharges, concentration at any one time depending much on the scale and intensity of upstream engineering activities. Suspended sediment concentrations exceeding 500 mg l^{-1} at well below the bankfull stage were measured on several occasions in the post-channelisation phase. Brookes (1987) documents enhanced suspended sediment loads downstream of 11 channel modification works, with levels attaining 40% greater than those upstream. Sedimentation was found to be greatest immediately below each scheme, declining with distance, whilst further downstream deposition was at a maximum in pools. Similar impacts have been documented by Sear et al. (1998) arising from the restoration of the lowland river Cole. Suspended sediment loads downstream of the restored channel were up to 150% in excess of input values to the restored reach. The evacuation of construction sediments has resulted in the development of silt benches and local gravel accumulations for up to 100 m downstream of the restoration site. Vegetation colonisation is anticipated to reduce sediment loads and with the restoration of overbank flood conveyance may reduce downstream sediment yields still further.

The impacts of these increased loads on river systems has not been well documented. Alabaster and Lloyd (1982) argue that good freshwater fisheries cannot be maintained above suspended sediment concentrations of 80 mg l^{-1}. Essery and Wilcock (1990) show that this threshold was exceeded about 8% of the time on the River Main prior to channelisation and about 20% of the time immediately following channelisation. Equally problematic is the duration of these impacts. The transformation of cobble-covered river beds into ones covered with silt is a matter of common observation after major engineering works. But, ten years after channelisation the silt deposits appear to be thinning on the River Main, presumably transferred periodically downstream towards Lough Neagh in successive floods. In contrast, Brookes (1987) observed rapid removal of construction sediments from the river bed following channel modification works.

3.6 Hydraulic and hydrological impacts of channel modifications

The alteration in channel form and the subsequent adjustment of the river channel promotes changes in the hydraulic and hydrological regime of modified channels. These range in scale from local scour and vortex shedding

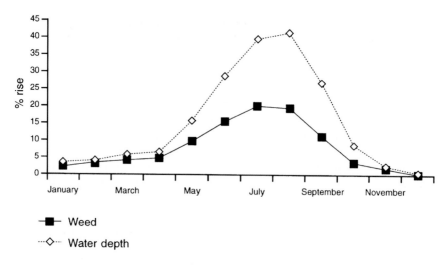

Figure 3.5 Change in water level in response to seasonal growth of aquatic vegetation.

around groynes or spurs, to major changes in hydraulic diversity associated with the simplification of morphology or the restoration of hydraulic diversity. The principal hydraulic components of channels are associated with the grain, form and floodplain roughness conventionally subsumed within a single roughness coefficient, e.g. Manning's 'n' or Darcy-Weisbach 'ff'. Simplification of channel morphology and boundary roughness effectively reduces flow resistance, resulting in more predictable and hydraulically efficient channels capable of transmitting much higher discharges for a given change in stage. Factors such as the annual growth of aquatic vegetation or the development of bars, pools and riffles can substantially alter the design conveyance of the modified channel, necessitating some form of maintenance. For example, Figure 3.5 shows the percentage rise in water levels associated with seasonal growth of aquatic vegetation for a small lowland stream. The roughness coefficient for the channel in June is much higher as a result of vegetation growth and results in a 40% increase in stage. Such relationships will vary depending on the type and amount of vegetation in the channel as well as the form of the channel cross section. Nevertheless, this basic principle results in the annual cutting of aquatic plants throughout many UK rivers, particularly in the intensively farmed lowlands where higher nutrient levels in stream water draining agricultural land may exacerbate weed growth.

The effects of channelisation on river hydrographs are well known. Bailey and Bree (1981), for example, demonstrated that flood peaks were 60% higher on rivers that had been arterially drained in comparison with rivers not so affected. Whilst as early as 1945, O'Kelly estimated that loss of flood storage resulting from embanking or dredging could increase downstream peak flows by 12–77% depending on the flood area eliminated. In general,

Table 3.5 Design capacity of arterial channels (return period of flood) and their hydrological effects on maximum and minimum flows (after Robinson, 1990).

	Increase in flows after scheme (%)		Channel design capacity (years)
	Annual maximum	7 day	
Ock (Oxfordshire)	14	10	10
Barlings Eau (Lincolnshire)	34	28	10
Witham (Lincolnshire)	74	57	50
Ewenny (Mid-Glamorgan)	88	53	100

arterial drainage schemes alter the flood response, leading to increased flood flows and shorter response times. This change is especially large for the biggest events, which are subject to the greatest amount of overbank storage and attenuation. Whilst it may be argued that every channelisation scheme will affect flows to a different extent, Robinson (1990) found a close relationship between the design flood capacity of a scheme and the magnitude of its hydrological effects (Table 3.5). Larger channels will contain higher flood flows within them and are more likely to tap shallow aquifers and draw down the water table. Arterial improvements reduce local flooding, but may simply move the problem further downstream. Robinson (1990) describes the impacts of a 1970s arterial drainage scheme on the River Witham, Lincolnshire. Between 1965 and 1978, over 50% of the main river network upstream of Lincoln was embanked, widened and deepened to protect adjacent farmland with a channel design capacity sufficient to contain the estimated 1:50 year flood (Figure 3.6). Unit hydrograph analysis identified a statistically significant decrease in the time to peak following channel modifications that Robinson attributed to loss of flood attenuation following elimination of floodplain storage. The observed mean annual flood increased from 15.6 before channel improvements to 27.2 $m^3 s^{-1}$ afterwards, a rise of over 70%. More recently, hydrological modelling of mean daily flows on catchments instrumented to monitor the effects of a channelisation scheme on the River Main in Northern Ireland was able to demonstrate systematically increased high flows and systematically reduced low flows over a two year period in comparison with an adjacent river on which channelisation had not been undertaken (Wilcock and Wilcock, 1995). The alleviation of low flows is an increasing feature of the hydrological management of modified rivers. Low flow impacts can be exacerbated where channels have been modified to convey high discharges, leading to lower than normal water depths, and absence of hydraulic diversity. Removal of riffles and runs, where these provide opportunities for aeration, can substantially reduce water quality during low flows. An increasing option in these circumstances is to alter

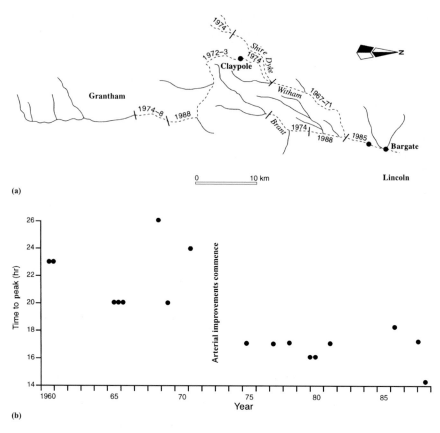

Figure 3.6 Impacts of channelisation on unit hydrograph, River Witham, Lincolnshire: (a) location and timing of River Witham channel works; (b) unit
hydrograph time to peak values at Claypole, for large events in the
period 1960–88. (After Robinson, 1990.)

channel form through narrowing bed width or reintroducing hydraulic diversity through restoration of riffle-pool sequences. The degree to which this
succeeds is still unknown and there is a clear need for specific research to
validate what is becoming a common management practice.

One of the major impacts of major channelisation schemes is loss of
floodplain wetlands. Attempts to quantify this process in terms of changes
to the streamflow into and out of an area of wetland have been attempted for
the River Main in Northern Ireland, the channelisation works for which
took place through an extended area of lowland raised bog and surrounding
fen, subject to regular flooding several times per year before channelisation.
The river channel before channelisation was sinuous, choked with sediment
and macrophytes, and occupied a wide flood plain. Inflow into this wetland
area, monitored automatically by conventional streamflow gauging equipment

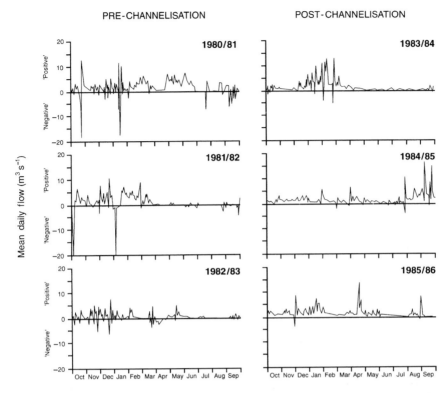

Figure 3.7 Estimated hydrographs of streamflow from floodplain wetlands in the upper River Main basin, Northern Ireland, before and after channelisation. The hydrographs presented here are derived as the daily streamflow at a gauge downstream of the wetlands minus the sum of flows on the three upstream tributaries. Negative flow days therefore represent periods of storage within the floodplain wetland when the sum of upstream inputs exceeds outflow. (After Essery and Wilcock, 1990.)

on three tributary streams (Figure 3.7), exceeded output on 344 days in a three-year period (31% of the time) prior to channelisation, mainly in winter. In the three years following channelisation, the number of such days was reduced to 40 (4% of the time) (Essery and Wilcock, 1990). No systematic ecological monitoring of the terrestrial ecology before and after channel modification has taken place in this area but much of the former, regularly-inundated, floodplain wetland has now been ditched, fenced, and reclaimed for livestock farming. In the unfenced areas, shrubs and trees appear to be slowly invading areas of former fen. The loss of lowland floodplain wetlands and associated functions of water, sediment and nutrient storage have profoundly affected the hydrology and geomorphology of UK river channels.

Hydrological and hydraulic impacts of river rehabilitation are potentially substantial and necessitate careful consideration during design and post

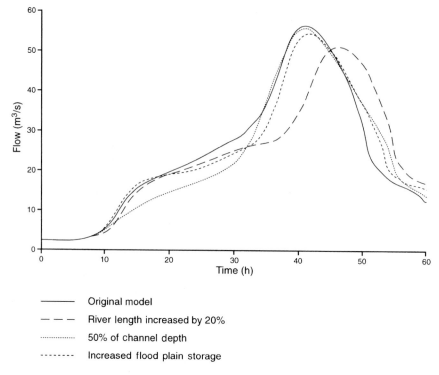

Original model
— — — River length increased by 20%
.... 50% of channel depth
- - - - - - - Increased flood plain storage

Figure 3.8 Impact of different restoration options on the 1 in 100 year hydrograph; River Cole, Oxfordshire.

project appraisal. The hydraulics of natural channels are complex and can vary spatially and with time. In practice, values for flow resistance change with increasing flow depth and throughout the year as vegetation grows and dies back. Furthermore, as rehabilitation schemes get older, the role of vegetation will change as arboreal species begin to interact with hydraulic and geomorphological processes across the flow range. More recently, there have been moves to bring rivers back into contact with their floodplains on a more regular basis. The change in hydraulics within a river channel, and the resulting hydrology of the floodplain are complex and remain difficult to predict, particularly when many floodplains have networks of underdrainage. Nevertheless, the re-introduction of floodplains for natural storage affords the greatest opportunity to manage catchment flood hydrology. This may be achieved by raising river bed levels and increasing flow resistance by recreat-ing natural channel features such as riffles and meanders. Computational modelling studies of different scales of rehabilitation for a mixed geology lowland catchment have indicated the positive flood defence benefits that might be achieved (Figure 3.8). Clearly this benefit is dependant on the type of catchment and the nature of the river network but as Figure 3.8 indicates headwater stream systems could be used to decrease flood peaks further down the catchment.

3.7 Prospects

At the end of the twentieth century, a reflection on the impact of channel modifications reveals a picture of a massively altered and heavily managed river network that nevertheless exhibits national variability in extent of impacts and uptake of rehabilitation and enhancement opportunities. Institutionally, it is arguable that in the past channel modifications, although undertaken by different organisations within any one catchment, were driven by the same motives and operated similar design principles. Today the picture is much more confused, being one in which a complex patchwork of organisations and individuals within any single catchment struggle to balance the legacy of existing channel modifications with new standards for environmental protection and enhancement, whilst being administered and funded in different ways within an environment that is itself changing. Truly integrated river management is not yet with us, and although steps have been made, it is clear that within the UK as a whole, radical changes in the management of land and water are still needed to deliver sustainable water management. Reversal of past channel modifications within this system are likely to remain opportunistic, and constrained to those administrative areas with vision and significant financial resources set aside for enhancements. Financially there is a clear need to modify existing cost:benefit methods to include the longer-term benefits of environmental enhancements. In addition, substantial re-education of riparian owners, farmers and authorities to the benefits of rehabilitation is needed, although just whose task this is remains uncertain.

On the ground, the management of river channels continues to be reach based, although the advent of Local Environment Agency Plans (LEAP) within England and Wales offers some hope for a more strategic delivery of channel modification. Similarly the monitoring of channel modifications is being made easier by the implementation of River Habitat Objectives, based on the River Habitat Survey, and which provides a national picture of the degree of channel modifications and, with time, the changes in the habitat quality brought about by enhancements and rehabilitation. As in all branches of environmental management, the ease of information gathering and transfer is accelerating so that twenty-first century river managers will have much more information on which to base decisions. There are however, notable omissions to this information since no progress has been made in the monitoring of sediment loads (suspended or bedload) in UK rivers, whilst in the science community the century ends with the effective decommissioning of the longest running catchment study to date, a mere twenty seven years.

The changes in UK river hydrology and geomorphology within the next century are going to be brought about by changes in the management of land set against development-based drivers such as the increasing need for new water resources and housing. Financial incentives to rehabilitate floodplains in order to derive maximum benefit for water quality and flood protection at the catchment scale will not come from modifications to water law. It is also clear that underpinning future channel management options must be robust scientific understanding of the interactions between ecological, hydrological and geomorphological processes and how these operate across scales within the river catchment so that people can more clearly recognise and mitigate impacts. It is probable that the era of environmental degradation through channel modifications has temporarily ceased. It is up to the water industry, environmental groups and science to deliver the sustainable management options and practices that will ensure that it does not recur.

References

Acreman, M.C., Elliott, C.R.N. and Gowing, I.M. (1996) Evaluation of the River Wey restoration project using the Physical Habitat Simulation (PHABSIM) model, in *Proceedings of the MAFF Conference of River and Coastal Engineers*, Keele, 3–5 July, 1996.

Alabaster, J.S. and Lloyd, R. (1982) *Water Quality Criteria For Freshwater Fish.* Butterworths, London.

Arnell, N.W. (1996) *Global Warming, River Flows and Water Resources.* Institute of Hydrology Water Series Publication. Wiley, Chichester.

Bailey, A.D. and Bree, T. (1981) The effect of improved land drainage on river flood flows, in *The Flood Studies Report, Five Years On*, Institution of Civil Engineers, London. pp. 131–42.

Brookes, A. (1987) River channel adjustments downstream of channelisation works in England and Wales, *Earth Surf. Proc. Landforms*, 12, 337–51.

Brookes, A. (1988) *Channelized Rivers: Perspectives for Environmental Management.* Wiley, Chichester.

Brookes, A. and Shields, F.D. (eds) (1996) *River Channel Restoration: Guidance for Sustainable River Management.* Wiley, Chichester.

Brookes, A., Gregory, K.J. and Dawson, F.D.H. (1983) An assessment of river channelization in England and Wales, *Sci. Total Environ.*, 27, 97–111.

Cruickshank, J.G. (ed.) (1997) *Soil and Environment: Northern Ireland.* Department of Agriculture for Northern Ireland and The Queen's University of Belfast, NI.

Dooge, J. (1987) Manning and Mulvany; River improvement in 19th century Ireland, in Garbrecht, G. (ed.), *Hydraulics and Hydraulics Research – a Historical Review.* Balkema, Rotterdam, pp. 173–83.

Dunbar, M.J., Elliott, C.R.N., Acreman, M.C. *et al.* (1997) *Combining Hydraulic Design and Environmental Impact Assessment for Flood Defence Schemes.* Report to MAFF FD0506 Institute of Hydrology.

Environment Agency (1998) *River Habitat Quality: the Physical Character of Rivers and Streams in the UK and Isle of Man.* Environment Agency, Bristol.

Essery, C.I. and Wilcock, D.N. (1990) The impact of channelization on the hydrology of the upper River Main, County Antrim, Northern Ireland – a long-term case study. *Regulated Rivers: Res. Manag.*, 5, 17–34.

Fitzsimmons, J. and Pimperton, A.J. (1990) Application of a computer-based system in the development of objective river maintenance. *J. Inst. Water Environ. Manag.*, 4, 154–62.

Gregory, K.J. (ed.) (1997) *The Fluvial Geomorphology of Great Britain.* Chapman & Hall, London.

Hutton, J.B.E. (1972) *Report of a Public Inquiry into the Proposed River Main Drainage Scheme.* HMSO, Belfast.

Newbold, C., Honnor, J. and Buckley, K. (1983) *Nature Conservation and the Management of Drainage Channels*, Nature Conservancy Council/Association of Drainage Authorities, Peterborough.

Newson, M.D. and Leeks, G.J.L. (1987) Transport processes at the catchment scale – a regional study of increasing sediment yields and its effects in Mid-Wales, UK, in Thorne, C.R., Bathurst, J.C. & Hey, R.D. (eds) *Sediment Transport in Gravel-bed Rivers.* Wiley, Chichester. pp. 187–223.

Newson, M.D. and Robinson, M. (1983) Effects of agricultural drainage on upland streamflow: case studies in mid-Wales, *J. Environ. Manag.*, Vol. 17, 333–48.

Newson, M.D. and Sear, D.A. (1994) *Sediment and Gravel Transport in Rivers.* National Rivers Authority, Bristol Project Report C5/384.

Northern Ireland Department of Agriculture (1992) *Tow River Ballycastle Flood Alleviation Scheme – Environmental Statement.*

Robinson, M. (1990) Impact of improved land drainage on river flows, *Institute of Hydrology Report No. 113.* Institute of Hydrology, Wallingford.

Robinson, M. and Rycroft, D.W. (1998) The impacts of drainage on streamflow, in van Schilfgaard, J. and Skaggs, R.W. (eds), *Agricultural Drainage*, Chap. 23. American Society of Agronomy, Madison, Wisconsin.

Sear, D.A. and Archer, D. (1998) The effects of gravel extraction on the stability of gravel-bed rivers: a case study from the Wooler Water, Northumberland, UK, in Klingeman, P., Beschta, R., Komar, P.D. and Milhous, R. (eds), *Gravel-bed Rivers in the Environment.* Water Research Council, Colorado, USA. pp. 415–32.

Sear, D.A., Armitage, P.D. and Dawson, F.D.H. (1999) Groundwater-dominated rivers, in Sear, D.A. & Armitage, P.D. (eds), *Groundwater-dominated Rivers.* Special Edition, *J. Hydrol. Process.* Wiley, Chichester.

Sear, D.A., Briggs, A.R. and Brookes, A. (1998) A preliminary analysis of the morphological adjustment within and downstream of a lowland river subject to river restoration. *Aquatic Conserv.*, 8(1), 167–84.

Sear, D.A., Newson, M.D. and Brookes, A. (1995) Sediment-related river maintenance: the role of fluvial geomorphology. *Earth Surf. Proc. Landforms.*

Swales, S. (1982) Environmental effects of river channel works used in land drainage improvements. *J. Environ. Manag.*, 14, 103–26.

Wilcock, D.N. and Essery, C.I. (1991) The impact of channelization on the River Main, County Antrim, Northern Ireland. *J. Environ. Manag.*, 32, 127–43.

Wilcock, D.N. and Wilcock, F.A. (1995) Modelling the hydrological impacts of channelization on streamflow characteristics in a Northern Ireland catchment, in Simonovic, S.P., Kundzewicz, Z., Rosberg, D. and Takeuchi, K., *Modelling and Management of Sustainable Basin-Scale Water Resource Systems*, International Association of Hydrological Sciences Publication No. 231, 41–48.

4

CAUSES OF CATCHMENT SCALE
HYDROLOGICAL CHANGES

Angela Gurnell and Geoff Petts

This chapter explores the hydrological consequences of water resource manipulation in the UK. It reviews the dominant artificial influences causing hydrological changes (such as groundwater abstraction, dams and inter-basin water transfers), and considers the spatial mismatch between water supplies and demand, using the history of water use and waste-water disposal in London and the River Thames as an example.

4.1 Background

Total freshwater resources in Britain are far in excess of demand but the balance between water resource availability and demand varies widely from one region to another. As a result, water management within and between catchments has been necessary in many areas and has resulted in considerable hydrological changes at the local, regional and national scales. Most rivers are regulated by dams and reservoirs (Plate 4.1) to support water abstractions for domestic, agricultural and industrial use, to control flooding, or to provide hydro-electric power. Early developments of water resources were intimately linked to problems caused by the disposal of the waste waters and other waste products. By the mid-nineteenth century, half of the population in Britain lived in towns. Many rivers were polluted by waste discharges, and problems of poor water quality were often exacerbated by the abstraction of naturally clean water that would otherwise have diluted polluted river water. Paradoxically, these problems were made progressively worse by the rapid advances in drainage of industrial and household wastes, motivated by public health concerns. The impact of the 'water closet' on downstream rivers was particularly dramatic (Sheail, 1988). Receiving streams became dominated in summer by effluent discharges. This legacy remains and in the West Midlands treated effluents form more than 70% of the dry weather flow in the River Tame.

Plate 4.1 Walshaw Dean reservoir, West Yorkshire (Mike Acreman).

4.1.1 'Natural' and modified river flows

As a result of far-reaching human influences on river flows, of the approximately 1000 UK gauging stations for which data are available for the period to 1990 (NERC, 1993), only 19% record 'natural' flows (35% in Scotland and 14% in England and Wales, with less than 6% in Anglian, Midland and Thames regions). These 'natural' flows are where there are no abstractions and discharges, or where variation due to abstractions and discharges are so limited that the gauged flow is considered to be within 10% of the natural flow at, or in excess of, the 95 percentile flow. Half of these stations receive runoff from drainage areas of less than 100 km^2 and only 10% have catchments of greater than 500 km^2. The largest remaining basins drained by 'natural' rivers are the Dorset Avon (1478 km^2) and the Yorkshire Swale (1363 km^2) in England and two rivers draining the Grampian Highlands of Scotland, the Dee (1370 km^2) and the Don (1273 km^2). However, even these rivers are being exploited and the pressures are intensifying (Gilvear, 1994).

The impact of the different causes of hydrological change is illustrated by considering the remaining stations which have artificial influences on their flow record. Of this total of 81% of gauging stations:

- 26% are affected by storage or impounding reservoirs;
- 16% are regulated through flow augmentation from surface water reservoirs and/or groundwater storage, including flows to meet the needs for power generation (this latter influences 11% of the stations in Scotland);
- 63% are affected by abstractions from surface water and from groundwater sources for public, industrial and agricultural water supplies (this includes over 70% of the stations in England and Wales);
- 33% are affected by significant outflows from sewage treatment works which augment river flows if the effluents originate from outside the catchment.

Over the past twenty years, the modification of the hydrological regime of British rivers has involved increasingly sophisticated techniques including the complex conjunctive use of reservoir storage, groundwater, river regulation, and inter-basin transfers. A good example of a conjunctive use system is operated by the North West region of the Environment Agency (Walker and Wyatt, 1989).

4.1.2   The regulated rivers of Britain

Figure 4.1(a) locates the major rivers of Great Britain that are regulated by surface reservoirs. Early reservoirs which impacted severely on river flows were mainly for public water supply. Of the reservoirs marked on Figure 4.1(a), the Vyrnwy dam in the Severn catchment impounded the largest reservoir in Europe at the time of its completion in 1891 and marked the beginning of large dam building and large inter-basin transfers in the UK (Sheail, 1988). Hydro-electric power generation is a relatively recent development which has had a major impact on river flows. Although the first significant plant was commissioned in 1896, it was the introduction of high-voltage electricity transmission in the early 1930s which made possible the large-scale use of hydro-power (Johnson, 1988). Particularly rapid growth followed the North of Scotland Hydroelectric Development Act of 1943.

Flow regulation can result from many hydrological changes other than the operation of surface reservoirs. Where rivers are groundwater fed (mainly in the lowland south and east of England), a reduction in summer flows has been a consequence of widespread groundwater abstraction. Although groundwater has a long history of use, large-scale groundwater exploitation is mainly a feature of the twentieth century. The flow regimes of some large rivers, such as the Spey in Scotland and the Nene and Wessex Stour in England, are significantly affected by surface and groundwater abstractions. Some, such as the Bristol Avon, have flow-augmentation from groundwater. Others, including the Bedford Ouse and Thames, are affected by abstractions and the operation of navigation weirs.

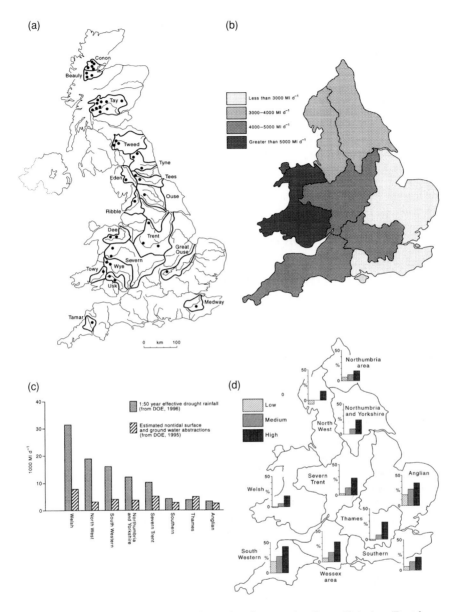

Figure 4.1 (a) Major reservoirs and regulated rivers in Great Britain. (Based on Petts, 1988; copyright © John Wiley and Sons Ltd. Reproduced with permission.) (b) Total groundwater and non-tidal surface water abstractions, 1993. (Information from DOE, 1995.) (c) Estimated non-tidal surface and groundwater abstractions compared with 1:50 year effective drought rainfall. (Information from DOE, 1996.) (d) Percentage increase in average public water supply demand 1991–2021 under three growth scenarios. (Based on NRA, 1994.)

Table 4.1 Source and purpose of total surface water and groundwater abstractions 1993 expressed as a percentage of total surface and groundwater abstractions (information from DOE, 1995).

Region (see footnote to table)	1	2	3	4	5	6	7	8
Surface abstractions (% of total abstractions)	50	88	88	76	45	85	66	98
Water supply	81	70	56	52	59	32	85	25
Agriculture and fish farming	6	12	6	1	32	47	19	3
Electricity supply and other industry	12	17	38	46	8	20	5	72

Regions: 1, Anglian; 2, Northumbria and Yorkshire; 3, North West; 4, Severn Trent; 5, Southern; 6, South Western; 7, Thames; 8, Welsh.

4.1.3 Patterns of water availability and demand

The spatial pattern of the estimated groundwater and non-tidal water abstractions in 1993 (DOE, 1995) is shown in Figure 4.1(b). The regional mismatch between estimated abstractions and the availability of effective rainfall in a drought of a severity which can be expected to occur once in fifty years (DOE, 1996) is illustrated in Figure 4.1(c). These two figures demonstrate the larger available water resources of the northwest and west of England and of Wales in comparison with the southeast and east of England. There are not only significant regional variations in the magnitude of total abstractions and the balance between surface and groundwater abstractions but also in the use of abstracted water (Table 4.1). Many parts of England and Wales have had to develop river basin management systems which ensure that water can be reused. For example, recycling of abstracted water takes place along the entire length of the River Thames. Figure 4.1(d) shows the forecasts of increasing demand for public water supply under three growth scenarios (National Rivers Authority, now the Environment Agency, NRA, 1994), indicating the need for further manipulation of hydrological processes in the future, although 'there is a strong possibility that demands can be managed to avoid the need for large scale water resources development over the next twenty years or so' (NRA, 1994, p. 7).

4.2 London and the River Thames: a history of hydrological change

The development of water supply within the River Thames basin, and particularly within London, illustrates a number of issues which have been repeated across Britain in association with the process of urbanisation, the main force driving hydrological change within British catchments.

Even before the period of major urban expansion, flows of British rivers and streams were altered for water power and to supply fish ponds. Figure 4.2(a) shows the considerable manipulation of the flow of the Thames and its tributaries as far back as the Domesday survey of 1086. Fish ponds and the systems of mill leets and races needed for milling flour and for other industrial purposes all had a regulating effect on river flows (Sheail, 1988). As towns were established, local sources were used for water supply. In the twelfth century, the city of London derived its entire water requirements from local streams and shallow wells. Early waste disposal systems were rudimentary. Natural water courses were usually used to remove wastes, frequently resulting in pollution of many of the rivers that were being used for water supply. Such conflicts induced further hydrological impacts which are exemplified below for London, but which were paralleled in urban centres throughout Britain:

- The first major hydrological impact on the Thames river system occurred when water demand from the growing population could no longer be met from local wells and streams, and so local transfers of water became necessary. For example, in 1236, pipes were installed to enable transfers of water from the River Tyburn to the City. Many similar conduits followed to supplement local surface and groundwater supplies.
- By the sixteenth century, such local transfers were insufficient to meet the growing demand and so in 1581 a second phase of hydrological impact commenced with the first large-scale pumping of water from the River Thames at London Bridge.
- Water demand continued to grow and so import of water over ever longer distances became necessary. For example, an Act of Parliament in 1606 established the 'New River Scheme': the import of spring water from near Ware in Hertfordshire along a 64 km aqueduct into London.
- During the seventeenth, eighteenth and nineteenth centuries, the problem of supplying a sufficient quantity of water was gradually overcome but water quality problems intensified as effluent from the growing population was discharged to local rivers. Most of London was still supplied through abstractions from the River Thames, which was now receiving large quantities of effluent. Although water companies endeavoured to improve the quality of water supplies (by changing sources, including groundwater abstractions, by constructing reservoirs, and by water filtration), outbreaks of cholera, notably in 1849 and 1866, bore witness to the poor quality of the water supply. The deteriorating state of the River Thames was highlighted by the hot summer of 1858, which became known as the 'Year of the Great Stink'.
- The construction and opening of new intercepting sewers to the north and south of the River Thames in 1865, carried effluent away from central London, transferring the pollution to downriver outfalls. Nevertheless, the River Pollution Prevention Commission, which was responsible for the River Pollution Prevention Act of 1876, suggested that the rivers Thames and Lee should be abandoned as sources of water and that water

(a)

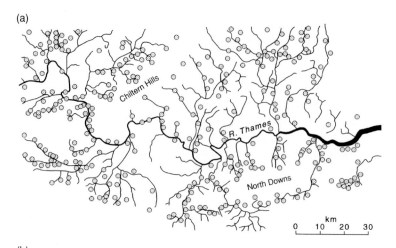

(b)

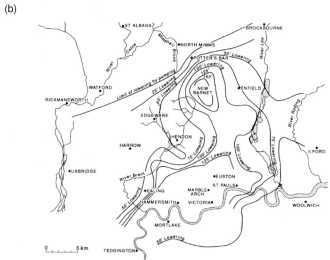

(c)

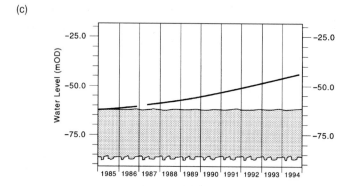

abstraction should revert to springs and wells within the London Basin (Walters, 1936).

- Groundwater abstractions developed rapidly and extensively during the late nineteenth and early twentieth centuries (Figure 4.2(b)). 'In the metropolitan area and in Central London there are about 750 wells and boreholes, from which several million gallons are pumped daily.... This has resulted in a general lowering of the water level under London, which in some places has reached 250 feet below sea-level. The rate of lowering in central London is from one to five feet per year' (Walters, 1936, p. 148). From borehole observations under Trafalgar Square, water levels in the confined Chalk and Upper Greensand aquifers fell by around 80 m between 1820 and 1940, a period when the aquifer was a major water source for London (DOE, 1995).

- Continued population growth and damage to sewers and sewage works during the Second World War led to continued decline in the quality of the tidal Thames. However, research on the dynamics of pollutant loads within the river coupled with post-war rationalisation and improvement of sewage treatment works have led to major improvements in water quality over the last thirty years. The improved water quality of the Thames and other surface water sources, and the development of surface reservoirs, has led to a reduction in groundwater abstraction. Furthermore, the improved quality of the Thames has supported some artificial recharge of the Chalk aquifer (e.g. Hawnt *et al.*, 1981). As a consequence of all of these recent changes, trends in water table levels have been reversed. For example, water table levels observed below Trafalgar Square stabilised in the 1940s and then began to rise. 'Over the last thirty years, the rise has averaged two metres a year and in parts of central London, now poses a threat to foundations and tunnels constructed while water levels were depressed' (DoE, 1995, p. 48; Figure 4.2(c)).

This complex interplay between population growth, increasing per capita water demand, effluent treatment and disposal, and the developments in technologies to support the use of both surface and subsurface water resources describe a complex history of hydrological change which provides a back drop to the following consideration of specific types of hydrological impact induced by major manipulations of water over the last 200 years.

Figure 4.2 (opposite) (a) Mills and fisheries on the River Thames and its tributaries, as recorded in 1086. (After Sheail, 1988; copyright © John Wiley and Sons Ltd. Reproduced with permission.) (b) Lowering of the water table under London 1878 to 1911. (After Walters, 1936.) (c) Water table levels below Trafalgar Square, 1985–1994 (bold line), in comparison with extreme monthly levels, 1845–1984 (stippled area). (After DOE, 1995. Material from Crown copyright work is adapted with permission of the Controller of Her Majesty's Stationary Office.)

4.3   River regulation and hydrological change over the last 200 years

Although hydrological change induced largely by human activities extends back over many centuries, major modifications of the hydrological regime of UK catchments have only occurred over the last 200 years. It is during this period that technological advances coupled with heavy domestic and industrial pressures on water resources have resulted in the construction of large surface reservoirs, the extensive development of groundwater resources, and the gradual implementation of inter-basin water transfers. Each of these three areas of development is discussed below.

4.3.1   Surface water development

Advances in water resources developments have involved the development of dams and storage reservoirs to reduce the natural variability of water yield within and between years, and to provide a controlled head of water for power generation. They also involved the development of transfer systems to redistribute water from supply to demand areas. At first these involved local transfers but long-distance, inter-basin transfers have become a feature of modern water management.

Dams and reservoirs

The construction of dams to create water storage reservoirs is a traditional water-management practice, dating back more than 5000 years. In the UK, the earliest attempts to organise and administer water supply on a collective basis have been attributed to the Romans by Binnie (1987) who cites, among other examples, the 1500 m aqueduct from the town of Wroxeter (the Roman Viroconium) to a dam site on a small brook. The oldest known dam in Britain, again dating from the Roman period, was not built for water supply but for water power at the Dolaucothi gold mines in Dyfed (Binnie, 1987). By 1086, the Domesday records refer to more than 5000 water-mills. However, the widespread and systematic development of impounding reservoirs for water supply dates from the late eighteenth century to meet the needs of growing urban centres whose local supplies were increasingly polluted. Thus, in the highly industrialized area of Lancashire and Yorkshire, small streams in the upland catchments of the southern Pennines were impounded to store part of the winter high flows to supplement supplies during the low-flow summer period. By 1936 nearly 200 small reservoirs had been constructed in the Pennines. Together they commanded a total catchment area of 1035 km^2 and had a combined capacity of about 280 million m^3 (Walters, 1936). The large number of reservoirs reflected an intensifying conflict between the need to abstract water from donor rivers to supply the growing towns and the need to maintain in-river flows to sustain the water power needs of the, often long-established, cotton and woollen mills (e.g. Petts, 1989).

The 15.5 million m³ Longendale Works, a series of five reservoirs on the River Etherow and two others in an adjacent valley, were built between 1848 and 1884 to supply water by gravity to Manchester. This marked the opening of the era of direct-supply impoundments but it was the completion of Vyrnwy dam in 1891 which initiated the modern phase of river regulation. The 60 million m³ Vyrnwy scheme represented a significant advancement in dam-building technology – the dam was the first to be over 40 m high – and in water distribution systems, creating the first major inter-basin transfer of water of over 100 km. The rate of large dam-building (structures over 15 m high) peaked in the 1960s (Beaumont, 1978). To demands for water-supply were added flood control and hydro-electric power production. Today, there are over 450 large dams in the UK, of which 80% are in upland areas, but only 15% have surface areas larger than 2 km². Petts (1988) isolated thirteen rivers from a list of all principal rivers in Britain on the basis of degree of regulation by dams and reservoirs (Figure 4.1(a)). This list is headed by three Scottish rivers, the Tay, Conon and Beauly, regulated for hydro-power production. In England, the most regulated river is the Tyne, and in Wales it is the Dee. Brianne dam on the River Towey in Wales is the highest at 91 m and Fannich reservoir on the Conon has the largest capacity (376 615 Ml).

Inter-basin water transfers

In the early eighteenth century, goods were transported by water whenever possible to minimise transport costs. A canal system was developed across England and Wales to improve the connectivity of internal water-borne transport. Between 1770 and 1830 over 2000 km of canals were constructed (Binnie, 1987). Figure 4.3(a) illustrates the complex network of canals that had been constructed by the end of the eighteenth century. Large quantities of water are required to maintain canal systems and many diversions of water from river systems were implemented and over 150 canal reservoirs were built to store water during the winter and to maintain canal levels in summer. Thus the canal network formed an early system of inter-basin transfer of water. Its potential is indicate by the NRA (1994) who propose some use of the canal network for water transfer as a part of their environmentally-sustainable water resources strategy (Figure 4.3(d)). Nevertheless, inter-basin transfers of water for water supply did not occur to a significant level until much later than the mid-nineteenth century because 'If a town endeavours to develop any resource in a valley other than that in which it is situated, even though quite near its borders, there may be great opposition; in fact it is generally a *sine qua non* that a town should reasonably develop the whole of its "own"

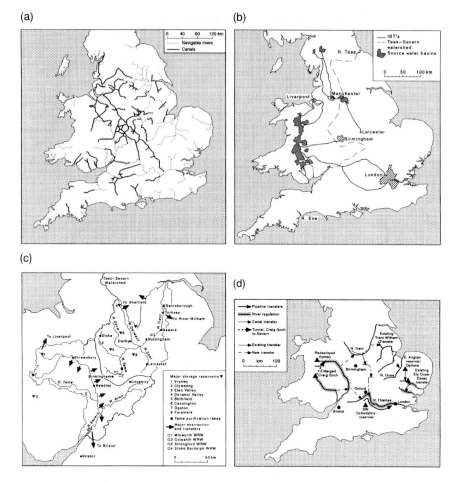

Figure 4.3 (a) The canal network of England and Wales in 1801. (After G.M. Binnie (1987), *Early Dam Builders in Britain*, Thomas Telford Ltd, London.) (b) Nineteenth-century plans for inter-basin transfers and their gathering grounds. (Reproduced from Petts and Wood (1996), with permission of Brewin Books, London.) (c) The Severn and Trent catchments showing major storage reservoirs, abstractions transfers and effluent discharges. (Reproduced from Petts and Wood (1996), with permission of Brewin Books, London.) (d) Strategic options as components of a National Water Resources Strategy. (After NRA, 1994.)

"natural" resources before promoting a Bill for a scheme elsewhere' (Walters, 1936, p. 107).

The first major inter-basin transfers occurred in association with reservoir development in Wales. Figure 4.3(b) illustrates the nineteenth century plans for inter-basin transfers to Merseyside, Birmingham and London and their

Welsh gathering grounds. The first scheme, completed in 1892 was the construction of the Vyrnwy reservoir and a 100 km pipeline which supplied the water to Liverpool. This was followed in 1904 by the implementation of a 118 km pipeline to Birmingham which transferred water from reservoirs in the Upper Wye catchment. The major schemes planned for the Lake District to supply Manchester involved increasing the capacity of two lakes, Thirlmere and Haweswater, and then transferring the water to Manchester. The Thirlmere scheme, initiated in 1879, involved 170 km of aqueduct, including tunnels, covered channel and piped sections. Although the long pipelines to London shown in Figure 4.3(b) were never constructed, inter-basin transfers have continued to develop both as shorter pipelines crossing catchment boundaries and as transfers resulting indirectly from the abstraction of water in one catchment and the release of effluent in a different catchment. For example, Figure 4.3(c) locates major storage reservoirs, abstractions, transfers and effluent discharges within the Severn and Trent catchments.

Development versus management

Since the Bills submitted to Parliament during the late nineteenth century to construct large reservoirs in the Vyrnwy and upper Wye valleys of Wales to supply Liverpool (1880) and Birmingham (1892), consideration has been given to providing compensation flows to meet a range of in-river needs, including fisheries and ecology. The Vyrnwy Bill created a precedent by allocating a discharge specifically for 'flushing purposes'. The Birmingham Bill addressed the needs of fisheries. Later, in 1919, a Bill to enlarge Haweswater included the setting of a guaranteed minimum flow with periodic freshets to improve conditions for fish breeding. Thus, a multi-purpose approach to river development was advanced in which concessionary modifications were made to the primary objective of water supply, flood control or hydro-electric power generation to meet other concerns. During the 1980s, this approach has been replaced by one which focuses on integrated management within which social and environmental values have a greater role in determining the ways in which we use our natural resources. This is well illustrated by the management of our groundwater resources.

4.3.2 Groundwater development

Groundwater development in Great Britain has a very long history. As early as the fourth century AD flourishing settlements on the higher chalk downs in Dorset were forced into decline by the difficulty of obtaining a reasonable water supply from water tables (Taylor, 1970) which were probably lowered by the combined effects of abstraction, land-drainage, and land-use change. However, the twentieth century has witnessed both advances in technology, which have allowed the sinking of ever deeper boreholes, and also increases in demand, particularly in the south and east of England, which have been reflected in great increases in the amount of groundwater abstracted. The three principal British aquifers are the Permo-Triassic Sandstones, the Jurassic Limestones and the Cretaceous Chalk. All of these underlie areas of lowland England, where population densities are high and, therefore, where the demand for water is also high. As a result, these aquifers are highly exploited with the percentage of average annual replenishment abstracted being 41%, 15% and 27%, respectively (Foster and Grey, 1997). Foster and Grey (1997, p. 193) attribute the great strategic significance of groundwater resources in England and Wales to three factors:

- They provide a relatively local, high-quality, low-cost source of public and private water-supplies, totalling 2500 million $m^3 y^{-1}$.
- They contribute more than 50% mains water supply over most of eastern, central and southern England.
- They maintain dry-weather flows in streams of significant environmental and/or amenity value and sustain numerous wetland habitats.

Unfortunately, conflict can arise between the first two and the third of these factors. In 1990 the NRA identified a priority list of 40 locations within England and Wales that were affected by low flows which were perceived to result from water abstraction. Subsequent investigations (NRA, 1993) revealed that 30 sites were in groundwater-fed catchments (seventeen primarily on Chalk, five on other types of Limestone, six on Triassic Sandstones, one fed from sand and gravel deposits, and one where the groundwater source was not specified) and that the main cause of the low flows in 24 of these sites was groundwater abstraction. Indeed, of the five rivers identified within the Thames basin, the equivalent of between 35% and 70% of the water that would be available to the river under average conditions is abstracted for public water supply.

The NRA (1993) reports progress towards devising solutions to these specific low flow problems. In many cases, groundwater abstraction has impacted severely on the river flow regime, but the problem is frequently seasonal and also results from local rather than regional groundwater use. Under such circumstances local solutions can be proposed including:

- reduced groundwater abstraction, particularly at times of low flow;
- flow augmentation, whereby mainly groundwater, but occasionally imported surface water, is added to the river to maintain flows whilst groundwater is still abstracted for water supply;
- relocation of abstraction boreholes to sites that are further from the river;
- recirculation of river water by pumping some from downstream to upstream sites to increase upstream flows;
- lining of the bed of some sections of river channel to prevent infiltration of river water.

Nevertheless, the pressure on water resources has resulted in more extensive and complex problems in some areas. The catchment of the River Darent in North Kent provides a good example of such severe problems. Groundwater has been progressively developed within the Darent catchment during the twentieth century. Furthermore, water consumed within the catchment is lost, because treated sewage is discharged to the River Thames and not back into the Darent. Dramatic reductions in river flows have resulted and sections of the river have dried up completely for prolonged periods since the early 1970s. 'The NRA believes that the restoration of a natural flow regime and habitat in the River Darent is incompatible with the needs of the water supply companies to abstract large quantities of water from the aquifers, particularly from points close to the river channel' (NRA, 1993, p. 29). The severity of the problem has resulted in the development of a major scheme for the Darent based upon the estimation of an environmentally acceptable flow regime. A wide range of options including reduction in the quantity of water abstracted, effluent reuse, new resource development and flow augmentation have been considered in developing the scheme.

4.4 Management of hydrological changes

As the maldistribution between water availability and water demand becomes an increasingly pressing problem in England and Wales, inter-basin transfers are seen as a major component of future strategies. Figure 4.3(d) presents the strategic options considered by the NRA (1994) in their consideration of an environmentally sustainable water resources development strategy for England and Wales to the year 2021. The figure illustrates four new transfers (Wye to Severn, Severn to Trent, Severn to Thames, Trent to Ely Ouse, Essex) which will link with existing transfers to implement the west to east movement of water that was planned in the nineteenth century (Figure 4.3(b)).

In the past, hydrological change has resulted in the loss of ecological integrity: the degradation of habitats, e.g. by siltation (Carling, 1988) and loss of connectivity with floodplains (Petts, 1996b), decline of fisheries (Mann, 1988) due to dams creating barriers to migration of salmonids and to habitat degradation, reduction in biodiversity (Boon, 1988), and loss of conservation value (Hellawell, 1988). The environmental problems caused by past water resource developments as well as those environmental issues raised by the demands for additional developments require a new approach in water management. The basis of modern river management is the determination of 'the flows which need to be protected to ensure the river can support the abstraction requirements placed upon it without compromising important ecosystems'.

The Environment Agency, formed on 1 April 1996 by the Environment Act 1995, invokes three principles for effective water resources planning:

sustainability, precaution and effective demand management. One principal aim is to manage water resources to achieve the right balance between the needs of abstractors and instream uses: water quality, navigation, recreation, fisheries and conservation. This requires the objective setting of flow targets – a Minimum Acceptable Flow (MAF). To aid water resource planning, a MAF concept was formally introduced in the 1963 Water Resources Act. In practice, no formal MAFs have been set but since 1963 it has been the law that where an application for an abstraction licence is made and

> the application is made at a time when no MAF for the inland waters in question has been determined . . . the Authority, in dealing with the application shall have regard to the considerations by reference to which . . . a MAF for those waters would fall to be determined.
> (Section 40, Water Resources Act 1991)

This requirement was transferred to the Agency by the Environment Act 1995. Consequently, the MAF concept is embedded in Agency practices. It is applied by setting Prescribed Flows attached to abstraction licences. A Prescribed Flow is 'the flow set at a river gauging station to protect downstream uses'. In practice, it is usually either a 'hands-off' flow/level below which abstractions must cease, allowing flows to decline naturally, or a 'maintained' flow/level, supported by artificial means, such as releases from reservoirs or support from groundwater. The Prescribed Flow usually includes two components: (1) an allocation to meet in-river needs (e.g. fisheries, conservation, water quality, navigation); and (2) an allocation to meet the needs of downstream abstractors. Prescribed Flows have been used to protect downstream interests giving due regard to the historic sequence in which licences were granted, with more recent licences having a higher Prescribed Flow than older ones (a process known as 'stacking'). This process ensures that at any time established 'uses' are protected and existing abstractions are not adversely affected by any subsequent licence. However, since the introduction of the MAF concept in 1963, there has developed a much stronger appreciation of environmental needs which now should be accounted for more explicitly in water resources management. Under Section 7 of the Environment Act 1995 (formerly Section 16 of the Water Resources Act 1991) the Agency has a duty when determining abstraction licences to exercise its powers such as to '. . . further the conservation and enhancement of natural beauty and the conservation of flora, fauna and geological or physiological features of special interest.' That duty extends, by virtue of Section 7(3), to each water undertaker's own functions.

Historically, in the UK, in-river needs were usually defined as the 95th percentile flow (Q95). This was adequate to protect rivers in most cases because only a small proportion of the available resource was actually abstracted and abstractions were allocated from the reliable baseflow component of the annual hydrograph. Over the past two decades in particular, increasing pressure on water resources and the increasing use of 'stacking' licenses has led not only to the greater exploitation of the reliable baseflow but also to abstractions during periods of higher flow. Consequently, the flows available for the environment have declined and the need has arisen to define environmental needs more precisely.

Current practice in some Environment Agency Regions recognises the need (1) to vary the level of flow required for environmental protection on different rivers (e.g. North East), (2) in some water resource schemes to have complex flow control rules to meet environmental, especially fisheries, objectives (e.g. the Roadford Scheme in South West Region), and (3) on some rivers to use hands-off flows to protect summer spates for fisheries (e.g. North West Region) or, on regulated rivers such as the Welsh Dee, to make special reservoir releases for migrating fish.

Recent scientific advances in the UK, USA and elsewhere (see review by Petts and Maddock, 1994) have developed from this simple consideration of minimum summer flows and spate-sparing for protecting the river environment, to recognise the importance of the full natural range of flows for sustaining healthy rivers. Thus, scientists are working towards defining Flow Regimes, which incorporate a range of ecological and environmental targets, to protect and enhance river ecosystems (Petts, 1996a; Petts *et al.*, 1998). These new approaches recognise that there is a sustainably exploitable water resource, whose quantity may vary between rivers and at different locations along a river, but that there is the need to protect low flows and to preserve the variability of flows, and that different rivers, and different sectors of rivers, including the estuary, have different flow needs. The new concept for managing our water resources will be 'acceptable hydrological change' defined in technological, economic, environmental and social terms.

References

Beaumont, P. (1978) Man's impact on river systems: a world-wide view, *Area*, I, 10, 38–41.

Binnie, G.M. (1987) *Early Dam Builders in Britain*. Thomas Telford, London.

Boon, P.J. (1988) The impact of river regulation on invertebrate communities in the UK, *Regulated Rivers: Res. Manag.*, 2, 389–411.

Carling, P.J. (1988) Channel changes and sediment transport in regulated UK rivers, *Regulated Rivers: Res. Manag.*, 2, 369–88.

DOE (1995) *Digest of Environmental Statistics No. 17, 1995*. HMSO, London.

DOE (1996) *Indicators of Sustainable Development for the UK*. HMSO, London.

Foster, S.S.D. and Grey, D.R.C. (1997) Groundwater Resources: Balancing perspectives on key issues affecting demand and supply, *J. CIWEM*, 11, 193–99.

Gilvear, D.J. (1994) River flow regulation, in Maitland, P.S., Boon, P.J. and McLusky, D.S. (eds), *The Freshwaters of Scotland: A National Resource of International Significance*. Wiley, Chichester. pp. 463–87.

Hawnt, R.J.E., Joseph, J.B. and Flavin, R.J. (1981) Experience with borehole recharge in the Lee Valley, *J. Inst. Water Eng. Sci.*, 35, 437–51.

Hellawell, J.M. (1988) River regulation and nature conservation, *Regulated Rivers: Res. Manag.*, 2, 425–44.

Johnson, F.G. (1988) Hydropower development on rivers in Scotland, *Regulated Rivers: Res. Manag.*, 2, 277–92.

Mann, R.H.K. (1988) Fish and fisheries of regulated rivers in the UK, *Regulated Rivers: Res. Manag.*, 2, 411–24.

NERC (1993) *Hydrological Data United Kingdom: Hydrometric Register and Statistics 1986–90*. Natural Environment Research Council, Institute of Hydrology, Wallingford.

NRA (1993) *Low Flows and Water Resources: Facts on the Top 40 Low Flow Rivers in England and Wales*. National Rivers Authority, Bristol.

NRA (1994) *Water: Nature's Precious Resource*. HMSO, London.

Petts, G.E. (1988) Regulated rivers in the United Kingdom, *Regulated Rivers: Res. Manag.*, 2, 201–20.

Petts, G.E. (1989) The regulation of the River Derwent, *East Midland Geographer*, 11, 2, 54–63.

Petts, G.E. (1996a) Allocating water to meet in-river needs, *Regulated Rivers: Res. Manag.*, 12, 353–65.

Petts, G.E. (1996b) Sustaining the ecological integrity of large floodplain rivers, in Anderson, M.D., Walling, D.E. and Bates, P. (eds), *Floodplain Processes*. Wiley, Chichester. pp. 535–51.

Petts, G.E. and Maddock, I. (1994) Flow allocation for in-river needs, in Petts, G.E. and Calow, P. (eds), *Rivers Handbook*, Vol. 2. Blackwell Scientific, Oxford, pp. 289–307.

Petts, G.E. and Wood, P. (1996) Inter-basin water transfers and regional water management, in Gerrard, A.J. and Slater, T.R. (eds), *Managing a Conurbation*. Brewin Books, Studley, Warwickshire.

Petts, G.E., Bickerton, M.A., Crawford, C., Lerner, D.N. and Evans, D. (1998) Flow management to sustain groundwater-dominated stream ecosystems, *Hydrol. Proc.* In press.

Sheail, J. (1988) River regulation in the United Kingdom: an historical perspective, *Regulated Rivers: Res. Manag.*, 2, 221–32.

Taylor, C (1970) *Dorset*. Hodder and Stoughton, London.

Walker, S. and Wyatt, T. (1989) The development and use of medium term policies for operation of a major regional water resource system, in *Proceedings of the British Hydrological Society, Second National Symposium*, 4.47–56.

Walters, R.C.S. (1936) *The Nation's Water Supply*. Ivor, Nicholson and Watson, London.

Section 2

EFFECTS

5

RIVER FLOWS

Terry Marsh, Andrew Black, Mike Acreman and
Craig Elliott

This chapter considers the spatial and temporal variations in flow patterns across the UK highlighting low flows, flood and the implications of changing flow regimes for aquatic ecology. It also explores the future management of UK rivers through the setting of river flow objectives.

5.1 Introduction

Carved into the landscape of the UK are almost 1500 discrete river basins draining to the sea through over 100 estuaries (Figure 5.1). These river systems comprise over 200 000 km of watercourses which act as major agents of landscape modification and, in turn, their characteristics are strongly influenced by the catchments through which they flow (Plate 5.1). The diversity of the UK in terms of its climate, topography, geology and land use makes for a rich variety of regime types and aquatic environments. Each river, from headwaters to the sea, presents a unique progression of different physical and biological habitats (NERC, 1990). Trends in runoff and variations in flow patterns affect every element in what are both familiar and complex freshwater systems.

Overall outflows from all the rivers of England and Wales broadly equate to that of the River Rhine. In a global context, UK rivers are mere streams, being characteristically short, shallow and subject to considerable man-made disturbance. Discharge rates typically range through several orders of magnitude and low flows tend to be very modest. For this reason in particular, UK rivers are especially sensitive to regime changes resulting from climatic variation or the net effect of anthropogenic factors.

There are now few UK rivers, certainly less than 15%, whose regime can be considered natural. The need to drain land, to protect it from flooding, to control the flow of water for water supply or hydropower, or to use the watercourses for navigation, fishing or recreation, have all imposed change to a greater or lesser degree. Changing the physical characteristics of the river

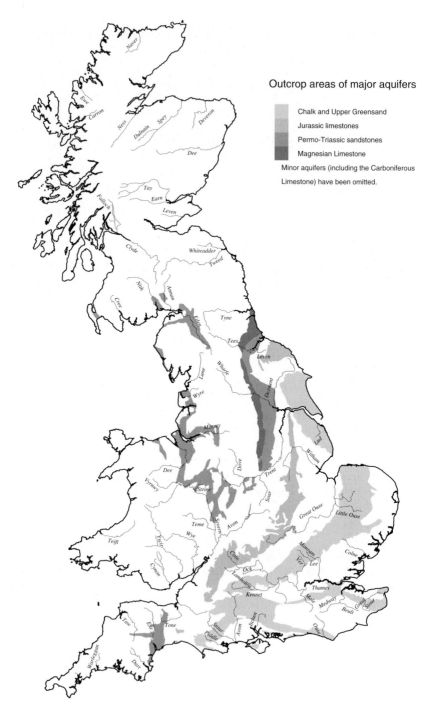

Figure 5.1 Rivers and aquifer outcrop areas.

(a)

(b)

Plate 5.1 River Kennet near Marlborough, Wiltshire: a lowland chalk steam dominated by groundwater showing large natural variations in flow (a) September 1997 (b) May 1998 (Mike Acreman).

– its width, depth and speed of flow – alters velocity and turbulence patterns and may have a range of effects on aquatic ecology by changing the availability and character of instream habitats. At the same time, land use changes influence both the catchment water balance and river flow patterns. Abstractions from rivers, and the groundwater which helps to sustain them, can lead to increased hydrological and environmental stress, particularly in the drier parts of the country, whilst floodplain development increases the vulnerability to out-of-bank flows. In response, water management has necessarily become more proactive in order to help minimise the risks associated with extreme rates of flow.

Rivers have always been an important part of Britain's history and culture. Today they are also required to make many less abstract contributions to our well-being. The contemporary management of river systems involves reconciling the often conflicting needs of man with those of the aquatic environment. Changing runoff patterns and flow characteristics, whether due to climate, land use or water utilisation, make achieving the right balance a major scientific challenge.

5.2 Evolution of UK drainage patterns and regime characteristics

Geological and geomorphological processes operating over tens of millions of years have bequeathed an exceptionally varied structural framework for UK river systems. Modern drainage patterns display a convincing imprint of the cycles of glaciation and deglaciation and the contemporary sea level changes during the Quaternary period (the last 1.6 million years). Their impact on valley form remains clearly evident and the associated mantle of superficial deposits provide an abundant source of material for redistribution by river action (Lewin, 1981). Palaeohydrological evidence confirm that erosion and sedimentation were more influential processes during a number of markedly different, and often more extreme, climatic episodes in the last one million years (Gregory *et al.*, 1987). Correspondingly, hydrological processes commonly differed greatly from those with which we are familiar today. For example, during the coldest periods periglacial conditions characterised the ice-free areas of the UK, river channels were commonly braided and maximum runoff rates tended to occur during the spring and short summer, the consequence of regimes dominated by meltwaters some of which could derive from areas beyond boundary of the topographic catchment (Barber and Coope, 1987).

The ready availability of gravel and alluvium facilitated changes in river morphology which could, under extreme flood conditions, be immediate and dramatic; for instance, the adoption of new river channels. By contrast, other forcing mechanisms, e.g. the regrading of the long profile in response to isostatic movement or sea level change, operate across very lengthy timespans.

River systems are always in natural transition and many of their characteristics may not reflect contemporary climatic conditions. Examples are found
in the ice-sculptured valleys of northern Britain and in southern England where
very modest streams drain extensive valley systems forged in the Chalk
outcrop during the glacial epochs (Dury, 1977).

The relatively new science of palaeohydrology is providing an increasing
wealth of material to help interpret current river networks and regimes in
the light of their complex climatic and geomorphological evolution (Brandon,
1996); the 20 000 years since the culmination of the last ice age have been
characterised by particular instability. By around 13 000 years before the
present, the climatic amelioration had produced conditions similar to those
of the modern era, but both the climate and landscape were continuing to
change. The initially impoverished post-glacial soils gradually matured to
allow the spread of thick forest cover (Barber and Twigger, 1987) which
itself affected the natural water balance. Very significant climatic perturbations
have continued through the last millenium with substantial temperature
variations; for instance, notably low temperatures characterised much of the
Little Ice Age (Lamb, 1977) whilst exceptionally mild conditions have been
typical of lengthy periods in the twentieth century.

Despite the complex interplay of many processes, the clear physical and
geological contrasts between upland and lowland Britain are reflected in
drainage and runoff patterns. The principal upland areas, mostly in the west,
are developed on the oldest rocks which are generally impermeable and
promote rapid runoff response. Lowland Britain, on the other hand, is founded
mostly on relatively young strata. The occurrence of extensive porous and
fractured rock layers interleaved between beds of impermeable clays is a
major feature of southern and eastern England (Figure 5.1). Such geological
contrasts make for marked regime differences. Figure 5.2 shows flow duration curves for six rivers draining catchments of particular geological character; in several of the catchments superficial deposits also influence the flow
patterns. Flow regimes can differ markedly even in adjacent catchments
subject to very similar climatic conditions. Figure 5.3 shows 1995 daily mean
flow hydrographs for the River Ock, which drains a largely impermeable
catchment in central southern England, and the neighbouring Lambourn, a
chalk stream. The greater responsiveness of the former and the muted flow
range of the latter are characteristic – groundwater is a major component in
the discharge of many lowland rivers. This baseflow makes for a relatively
stable flow regime and helps sustain flows through extended dry periods.

The great majority of UK rivers are perennial but exhibit a distinct
seasonal flow pattern. This is not a direct reflection of rainfall distribution
through the year. On average, precipitation is relatively evenly distributed
but with a clear tendency towards an autumn and early winter maximum in
the more maritime west. Principally it is evaporation losses, around 70%
of which are concentrated in the summer half-year, that impose a marked

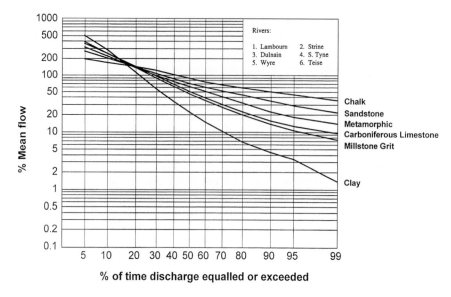

Figure 5.2 Flow duration curves for rivers draining catchments of differing geology.

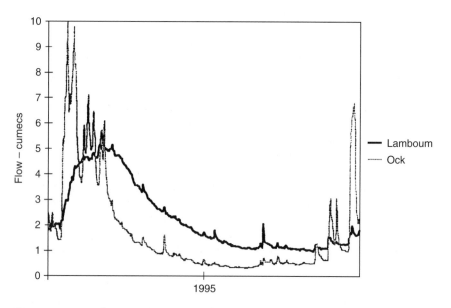

Figure 5.3 1995 flows in the River Ock and River Lambourn.

seasonality on flow regimes. In the mostly impermeable northern and western catchments where, lochs and lakes aside, natural storage is limited, occasional depressed winter flows result from frozen catchments but minimum flows normally occur in June or July. By contrast, in eastern rivers fed mainly by

106

groundwater, annual minima generally occur in the autumn, or even the early winter for streams which are entirely spring-fed. When, as during 1988–92 in much of eastern England, a succession of relatively dry winters provide only limited groundwater replenishment, river flows can remain depressed for very extended periods (Marsh *et al.*, 1994). Such episodes are accompanied by the failure of springs and the loss, albeit temporary, of aquatic habitat.

Floods and droughts in the UK do not pose the widespread threat to lives and livelihoods that they represent in much of the world. Nonetheless, in a densely populated country where there is a fine balance between water supply and demand in some regions, impacts such as water use restrictions can be inconvenient. Where development on the floodplain allows rivers an increasingly restricted access to their natural high flow province, levels of disturbance and financial loss associated with major flood events can be very large. For example, the insurance claims resulting from the Easter floods in 1998, which were most severe in the English Midlands, were of the order of £400–£500 million (Bye and Horner, 1998).

Although the periodic inundation of floodplains is a natural process, the generally irregular and infrequent nature of flooding creates the potential for locally unexpected problems, and justifies the need for scientifically-based programmes of flood hazard assessment and defence to be maintained. A range of flood generating mechanisms are important across the UK, often acting in tandem in individual rivers. Frontal rainfall is the dominant cause of flooding across the UK, with occlusions often leading to the highest rainfall intensities. Large catchments such as the Thames and Severn respond to heavy rainfall, often over several days and extending across many hundreds of square kilometres. Much more localised but intense precipitation events have been responsible for extreme flows in smaller catchments, such as occurred at Lynmouth with disastrous effect in 1952 (Dobbie and Wolf, 1953) and in Calderdale in 1989 when the erosive power of exceptionally rare storms was convincingly demonstrated (Acreman, 1989). Summer convective storms are often responsible for local flooding in urban and rural areas. Especially damaging floods, such as those which afflicted southern Britain in the summer of 1968, can result from a combination of heavy frontal rainfall and intense downpours associated with embedded thunder cells.

Snowmelt has been a major contributory factor to a number of notable flood events in northern Britain but is less important in the south; this is particularly true of the period since the early 1980s. However, in combination with heavy rainfall, snowmelt can cause major and widespread flooding such as occurred during the winters of 1979–81. In March 1947, the most extensive flooding this century followed the passage of a rain-bearing frontal system which generated a rapid thaw over still frozen ground. Floodplain inundations were very extensive. The Severn, Thames and Trent basins were badly affected. Flooding was especially severe in Nottingham and other

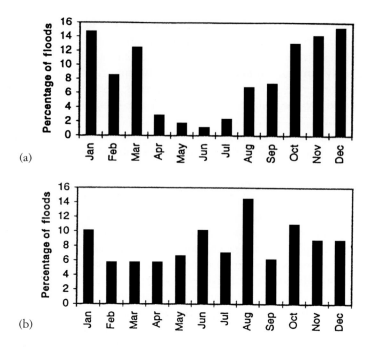

Figure 5.4 Monthly frequency of notable flows in the (a) River Falloch at Glen Falloch and (b) River Quaggy at Manor House Gardens.

towns and cities in the Midlands (Newson, 1975). Occasionally snowmelt alone can generate exceptional runoff rates. In 1963 snowmelt caused the highest known flood flows in the Wansbeck (Northumberland) after a severe winter freeze, despite the absence of any significant rainfall (Archer, 1992).

Characterising flood regimes, given the diversity of generating mechanisms and variety of catchment types, is a difficult task. Catchment geology and relief are important controls, spates being more common in mountainous, impermeable catchments than in rivers draining more subdued permeable catchments. One current area of research making a valuable contribution focuses on the seasonality of flooding; often flood generating mechanisms are associated with a particular seasonal signature. Black and Werritty (1997) describe the diversity of flood seasonality present amongst Scottish rivers, identifying links with soil moisture, rainfall seasonality, snowmelt and other factors. In most regions, flooding is mainly a winter phenomenon; soil moisture deficits substantially mitigate the flood risk in most catchments during the summer. Bayliss and Jones (1993) have observed a general pattern of mean flood date occurring later in the autumn/winter with distance from west to east across Britain. For the majority of western catchments, over 70% of floods occur during the October–March period (Figure 5.4). However, in urban catchments, especially those in the southeast, convectional storms in the summer and early autumn can be the primary cause of floods. This is reflected in the monthly distribution of flood events for the River Quaggy in south London. August is the modal month for exceptional flows. Notable late summer/early autumn storms can also produce severe flooding in eastern Scotland (Black and Werrity, 1997). A flood of this type produced the highest recorded runoff rate for the UK, 2241 $m^3 s^{-1}$ on the River Findhorn in August 1970. Comparable events, but of lesser magnitude, occurred in the same area in July 1956 (Green, 1971) and July/August 1997.

5.3 The impact of man on flow regimes

Disturbance of natural flow regimes is not a new phenomenon. With the birth of agriculture and population growth, man began to exercise a hydro-logical influence. This was imperceptible initially but in exploiting rivers for defence and water supply purposes, and catchments for food production, man became an increasingly active player. Over the last 1000 years, pasture improvements (the drainage of the lowlands especially), widespread defor-estation and urbanisation, all began to affect river systems in the UK (Higgs, 1987). Major land drainage works began in East Anglia during the sixteenth century and the New River has been augmenting London's water supply since 1613. Generally, however, up until the eighteenth century almost all water needs were satisfied from local sources. Regional water transfers, via aqueducts, were pioneered by Victorian engineers to satisfy growing urban and industrial demands. Thereafter regional and within-basin linkages de-veloped widely. In England and Wales, there are now about 50 000 licences for the impoundment and abstraction of water and overall demand is over twenty times that of a century ago. Today, the inter-linking of regional and local sources is commonplace and the regulation of rivers for water supply purposes is widely practised.

Floodplains have always attracted human settlement and currently around seven million people in the UK live in areas prone to regular natural flooding. Urbanisation and the resulting increased cover of impermeable surfaces has been shown to increase catchment responsiveness ('flashiness'), e.g. in new town catchments such as Harlow, with effects proportionately greatest for low return period floods (Hollis, 1975). The construction of reservoirs, retention ponds, flood banks and flood relief channels has moderated the risk in many vulnerable districts; around 36 000 km of flood defences now protect more than 1 200 000 ha throughout England and Wales. An examination of 4500 sites undertaken as part of The Environment Agency's River Habitat Survey (Raven et al., 1996) showed that less than 10% of sample river reaches were free from structural modification of channel and banks. More environment-ally sympathetic means of accommodating flood flows, e.g. on the River Skerne in Cleveland and the River Cole in Oxfordshire (Chapter 3), are

Table 5.1 Great Britain: water resources.

	Annual rainfall (1961–90 average) (mm)	Annual runoff (long term average) (mm)	Estimated abstractions as % of runoff
Great Britain	1080	660	7.5 (est.)
Scotland	1436	1040	1.1[a]
England and Wales	895	450	17.3
Regions:			
North West	1201	800	9.5
North East	834	435	12.7
Midland	754	340	21.0
Anglian	596	165	18.2
Thames	688	240	54.7
Southern	778	310	31.3
South West	1019	570	14.3
Welsh	1313	865	14.3

Source: National Water Archive and Digest of Environmental Statistics (HMSO).
[a] Public water supply abstractions only.

beginning to supersede the traditional straightening and deepening of channels. Increasingly, water management is also focusing on the alleviation and mitigation of artificially induced low flows. Flow augmentation, often using groundwater or sewage effluent, which may be diverted across catchment divides, is undertaken in a substantial number of catchments, mostly where runoff rates have declined to a degree which impacts on downstream water use or causes ecological stress.

Despite the ubiquity of man's influence on flow regimes, a broad distinction between upland and lowland catchments can readily be made. The former tend to be wet, have sparse settlement and are conceived to be 'natural', although land use impacts on flow patterns can be significant and the associated relief affords opportunities for natural or artificial impoundments to exploit the abundant rainfall in strategic reservoir systems. In addition, substantial water transfers may occur across catchment divides particularly in Scotland where runoff from around 20% of the country is utilised in hydro-electric power generation. Lowland Britain is much drier and the UK's lowest rates of runoff tend to coincide with concentrations of population, commercial activity and intensive farming. Thus, water demand, like rainfall, is very unevenly distributed across the UK, creating large regional differences in the vulnerability of rivers to the effect of abstraction (Table 5.1). In Scotland, public water supply abstractions constitute only around 1% of the total runoff whereas in the Thames region the corresponding figure exceeds 50%. Because of multiple reuse within the Thames basin, this figure needs to be treated with caution but it does signify very heavy utilisation of the available resource.

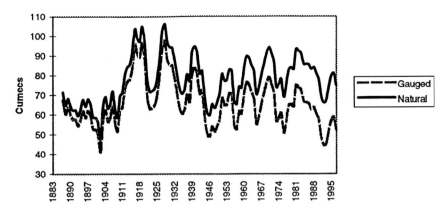

Figure 5.5 Gauged and naturalised flows (5-year running mean) for the River Thames at Kingston/Teddington 1883–1997.

Artificial influences can range from the gross disturbance resulting from the construction of a major water supply or flood retention reservoir, to the much more subtle effects of land use change and land drainage – examined, for example, in long-term hydrological monitoring programmes undertaken in the Plynlimon (Kirby *et al.*, 1991) and Coalburn (Robinson, 1998) catchments. Within individual catchments, different influences may reinforce or moderate the overall effect on the flow regime; for instance, the construction of reservoirs and ornamental lakes tends to dampen flow variability whilst the extension of arterial drainage can increase catchment stream flow peaks (Robinson, 1990). The impact of artificial influences in small and relatively homogenous catchments is reasonably well documented. At the basin and regional scale, however, the multiplicity of effects, which may change diurnally, seasonally or over longer timeframes, can make for a tenuous relationship between apparent causes and apparent effects. The quantification of such relationships is complicated by the large natural variability of river flows.

Man's impact on runoff can be readily demonstrated where long records of abstractions (and returns) allow the adjustment of gauged flows to produce a naturalised runoff series. The Thames basin provides a revealing case study. Routine flow measurement began at Teddington Weir, effectively the tidal limit, in 1883. Over the ensuing 115 years, gauged flows have declined substantially (Figure 5.5), emphasised by the new annual minimum gauged flow established in 1997. The character of the Thames, and the catchment it drains, has changed greatly from the nineteenth century when milling was very common, water meadows covered much of the floodplain and sewage effluent was only a minor flow component above the tidal limit. Notwithstanding these changes, the decline in flows primarily reflects the growth in abstraction to meet London's water demand. Today, abstractions in the lower reaches are 10 times those of the 1890s and, on average, exceed the mean gauged flow for August.

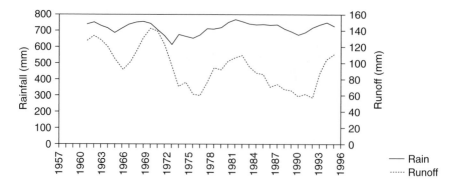

Figure 5.6 Annual rainfall and runoff (5-year running mean) for the River Ver 1962–96.

Low summer flows into the Thames Tideway can create water quality and occasional navigation problems but the steady improvement in quality in the lower Thames, and the return of the Salmon, demonstrates how skilled water management can allow heavy exploitation and a healthy aquatic environment to coexist. In a number of the Thames tributaries, however, a proportionally similar abstraction rate (mostly from groundwater sources) has produced a contraction in the stream network, a decline in headwater flows and a considerable loss of aquatic habitat. Although aquatic species, to some extent, adapt to natural variations in flow and occasional droughts, reducing the flow by abstraction means that critical thresholds are crossed more often, putting increased pressure on aquatic ecosystems.

The development of a method for estimating low flow statistics at ungauged sites (Gustard *et al.*, 1992) provided a means of broadly assessing the proportion of the stream network significantly affected by abstraction in the UK. In this study, all gauged catchments were classified according to the degree of artificial influence on low flows. Essentially the assessment procedure identified those catchments with a biased value of Q_{95} due to a seasonal net loss (abstraction) or gain (effluent return) to the catchment. For approximately one third of the 1366 catchments classified the gauged Q_{95}/mean flow ratio differed by more than 50% from the corresponding estimated natural ratio. Using a more subjective approach, the Biodiversity Challenge Group (BCG) identified 78 rivers perceived to be currently at risk from overabstraction (BCG, 1996).

The problem of artificially induced low flows is exemplified in the Chilterns. In the headwaters of the River Ver above St Albans, abstraction rates increased steadily from the turn of the century and, by the early 1990s, the overall abstraction was the equivalent of 30–40% of the natural replenishment to the groundwater which sustains the river. Year-on-year variability in runoff for the Ver can be large but, entering the 1990s, average flow rates were considerably below those of 40 years previously (Figure 5.6). Environmental concern triggered by such circumstances was the stimulus for the Alleviation of Low Flow (ALF) programme instigated by the National Rivers Authority (NRA) in 1991 (NRA, 1993) – the programme is now operated by the Environment Agency. Particular attention focuses on the actual or potential difficulties resulting from heavy abstraction from rivers or the groundwaters which feed them.

Local circumstances determine the best alleviation options. These may involve reduced or substituted abstraction, low-flow augmentation or bed lining to prevent leakage where the water-table is below the channel level. The ALF programme, which embraced 40 rivers and around 335 km of watercourses by 1995, has made an important contribution to rehabilitation of many valued streams. In the Ver, heavily reduced abstraction rates contributed to the steep recovery in flows following the 1988–92 drought. However, the degree of communal lobbying to help galvanise remedial action on rivers with much depleted flows varies greatly around the country and in some rivers, e.g. those draining parts of the Permo-Triassic sandstones outcrop of the Midlands, significantly reduced runoff has characterised much of the last 100 years.

The impact of man, and water management especially, on river systems has become very pervasive. Increased public expectations regarding the ecological and amenity status of watercourses and the further significant land use change in prospect (urbanisation and afforestation in the lowlands especially) may be expected to intensify the anthropogenic pressure. Such forcing mechanisms operate within a broader frame of institutional, economic and cultural mechanisms – for example the UK legislative framework and the policies of the European Union – which, together, will help shape the future evolution of UK river systems.

5.4 Trends in runoff: the instrumented period

The UK is fortunate in having river flow data extending back to the first half of the nineteenth century (Marsh, 1996a). This legacy is augmented by less formalised hydrometric information, as well as rainfall data, which shed light on earlier hydrological conditions. Reconstructed river flow series, normally based on rainfall records which in a few catchments begin in the seventeenth century, capture the broad characteristics of regime variation (Jones *et al.*, 1984). Such series are especially valuable in considering the frequency of drought conditions given current patterns of land use and water management. They are less suited to the quantification of actual trends or changes in regime. Long series of gauged flows are to be preferred but it should be recognised that most long river flow records are not homogeneous and many may not be representative. Considerable curatorial and analytical skill is normally required to properly identify trends in runoff patterns or regime types.

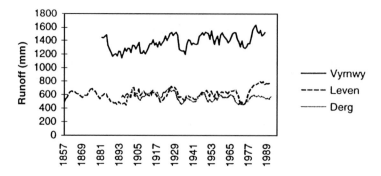

Figure 5.7 Long runoff records (5-year running mean), British Isles.

Notwithstanding the variable quality of historical data, most long runoff records for maritime western Europe are, like the corresponding rainfall series, typified by perturbations about a relatively stable mean (Green *et al.*, 1996). Figure 5.7 illustrates time series for Loch Leven (Scotland), Vrynwy Reservoir (Wales) and the River Derg (Ireland). However, periods of above, or below, average runoff can be very protracted; for example, in southern England low runoff rates were common during the thirty years from 1880 and in the 1940s. Both variability and persistence can be much greater in large parts of the world but the variations in the prevalence of wet, or particularly dry, episodes in the UK dictates caution in the interpretation of runoff trends. Apparently compelling trends exhibited by gauged flow data for a particular river or region may be revealed as far less convincing when examined in a wider time frame.

Examinations of lengthy river flow series may reveal inconsistencies or discontinuities which themselves may suggest or reinforce any apparent trend. Detailed analyses can often demonstrate that these reflect changing methods of flow measurement, variations in the hydrometric performance of the gauging station, or inconsistencies in the derivation and use of stage-discharge relationships. Artifacts can be particularly influential in the extreme flow ranges. On a number of the larger lowland rivers in England, for instance, many weirs installed primarily for navigation purposes were – relative to their modern counterparts – inadequate for accurate flow measurement and only broad comparisons with contemporary flows are possible. Changes in stage-discharge relationships can introduce step changes in a data series. This is a particularly important factor with regard to flood flows (Robson and Reed, 1996). Other factors can, however, be influential. A monthly sequence of flows for 1841–97, derived from a count of lockages of water fed from the Wendover Arm to the main Grand Union Canal, provides a unique insight into runoff variability in central England during the Victorian era (Anon., 1995). However, from around 1880, serious leakage from the Wendover Arm into the underlying chalk, resulted in a substantial underestimation of

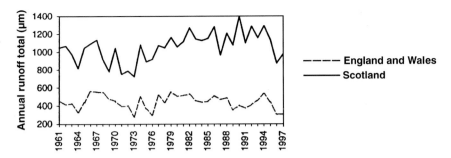

Figure 5.8 Annual runoff estimates for England and Wales and for Scotland.

runoff and an apparent decline in average flows. The need for critical reviews of historical series is not confined to little-used or recently unearthed datasets. A major rainfall/runoff modelling exercise using flows for the Thames at Teddington revealed that the long suspected underestimation of pre-1950 low flows – attributable mostly to leakage through the many gates and sluices which formed part of the Teddington weir complex – is of a greater magnitude than previously thought (Littlewood and Marsh, 1996).

On the basis of runoff data for a representative network of catchments, it is clear that there is little or no overall trend in runoff from the England and Wales as a whole over the last forty years (Anon., 1995) (Figure 5.8). For Scotland though, and particularly for rivers draining from the Highlands, runoff increased considerably over the 20 years to the mid-1990s (Curran and Robertson, 1991; Black, 1996). Changes in the frequency and magnitude of high and low flows are generally of greater significance than variation in the mean. Limited systematic examination of trends in low flows over the last 100 years has been undertaken, although shorter time series for individual catchments point to large regional contrasts and the effect of heavy abstraction rates is well documented. At the other extreme of the flow range, most land use change has an adverse effect on flood frequency but flood alleviation measures constitute a countervailing influence in many catchments (Institute of Hydrology, 1999). For the Thames, Crooks (1994), using recorded levels at locks, found a relatively stable rate of floods exceeding bankfull over the last 100 years although channel dredging and flood alleviation measures had resulted in a localised decline in peak levels. Robson *et al.* (1998) found no significant trend in a large UK flood database of annual peaks-over-threshold series extending to the 1980s and annual maximum series extending to the 1990s, although systematic fluctuations and strong climatic links are noted. A related study (Robson and Reed, 1996) demonstrated that in the minority of catchments where strong non-stationarity in the flood records has been identified, most trends are positive.

The need to capitalise more fully on historical and proxy flood data is widely recognised and advances in information technology are providing wider access, and more powerful searching facilities. Such initiatives as the British Chronology of Hydrological Events web site launched by the British Hydrological Society in 1998 will help to identify and collate valuable material which, despite its often incomplete nature and variable quality, provides potential insights into long-term hydrological variability. At present, however, our understanding of flow regimes and their variation through time is heavily influenced by data gathered in the post-1960 era. Rapid growth in the gauging station network followed the Water Resources Act of 1963 but even today the average length of flow record remains under 25 years. Most modern water resource systems, and most engineering design procedures, are based in large part on flow variability over the 30–40 years ending in the mid-1980s. Doubts persist as to whether the full range of historical variability is effectively captured in this timespan. It may have been a relatively quiescent period. Considerably more data acquisition, appraisal and analysis will be required to properly place it in a longer-term hydrological context.

5.5   Characteristics of the recent past

5.5.1   *Runoff*

The results of a number of investigations into the effect of climate change on flow regimes in the UK have suggested that the impacts may be substantial (Arnell and Reynard, 1996; CCIRG, 1996), although uncertainty bands are wide and there are considerable differences between individual scenarios. Given the modest size of UK river basins, and the scale and complexity of the world climate system, this is unsurprising. Certainly, the sensitivity of UK rivers to the combined effect of changes in rainfall and evaporative demands has been underlined during successive droughts in the 1990s; a 15% reduction in rainfall in East Anglia in 1988–92, for example, produced a 50% reduction in runoff in some catchments (Marsh *et al.*, 1994). However, UK climate change scenarios will only achieve full credibility when they can be reconciled, through time, with trends and regime changes identified from direct field measurement.

The recent past has been unusual in climatic terms. Although some historical similarities between the volatile rainfall patterns of the 1990s and those experienced in the 1930–50 period and in the 1850s may be recognised, once account is taken of the elevated temperatures (of the post-1987 period especially), there is no close modern parallel to the conditions recently experienced. Winters since the early 1980s have generally been very mild but with wide variations in rainfall totals, summers have typically been drier and hotter than their precursors (Marsh, 1996b). This tendency towards a more continental partitioning of rainfall between winter and summer has produced a general exaggeration in seasonal flow contrasts across the UK (Anon., 1996). Contemporary rainfall patterns have been unusual in a spatial as well

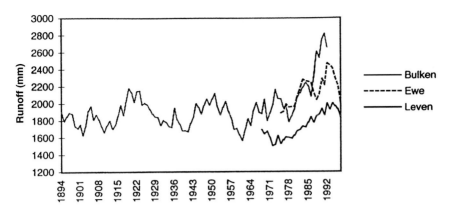

Figure 5.9 Post-1970 runoff (5-year running mean) increase for rivers in Scotland and Norway.

as a temporal sense. Over the 20 years to the mid-1990s, a reinforcement of the north–west to south–east rainfall gradient across Britain may be recognised, a pattern driven by increases in the frequency of westerly airflows (Mayes, 1995; Foster *et al.*, 1997), and also reflected in north-western Europe (Green and Marsh, 1997). One consequence is illustrated in Figure 5.9 which shows a steep increase in runoff from the Rivers Ewe and Leven, both of which drain from the Scottish Highlands. The significance of this trend is somewhat exaggerated by the depressed runoff rates in the 1971–76 period but it is principally the result of the unprecedented recent wetness in Scotland, the winter periods especially. A common synoptic backcloth is suggested by the lengthy flow record for the River Bulken in Norway, where runoff recorded over the 1985–95 period is well outside the range of previous variation.

For the UK as a whole, runoff over the 1988–97 period was very close to the long-term average but, at the regional scale, spatial variations were large, with some eastern catchments registering below 80% of the long term average. Interactions between the unusual rainfall patterns and high evaporative demands have produced atypical soil moisture conditions which contributed to large, and sustained, departures from average flow rates. Spate conditions were very common in western Scotland, the Highlands especially, whereas the English lowlands were afflicted by several notably protracted low flow episodes. Many unprecedented accumulated runoff totals in the 12–36 months time frames have been established in the 1990s although only a proportion of the outstanding minimum flows registered towards the end of the 1975/ 76 drought have been eclipsed (Marsh and Lees, 1998). The remarkable intensity of this event is exemplified on the Thames where, following the driest 16-month sequence in the England and Wales rainfall series, flows at Teddington ceased for the first time in recorded history; over wide areas, the return periods ascribed to the late-summer flows in 1976 exceeded 50 years

Table 5.2 Ranked annual 90-day minimum flows.

	South Tyne 62–97	Lud 68–97	Trent 58–97	Kennet 61–97	Mimram 52–97	Great Stour 64–97	Exe 56–97	Cynon 57–97	Eden 67–97
1	*1995*	*1991*	1976	1976	1976	*1990*	1976	1984	*1995*
2	1976	*1996*	1959	*1997*	1973	*1997*	1975	1976	*1996*
3	1996	1976	*1990*	*1990*	*1997*	*1996*	1984	1989	1976
4	1989	1989	1989	1996	1991	1976	1995	*1995*	1972
5	*1994*	1995	1996	1991	1992	*1989*	1978	1975	*1989*
6	1974	*1992*	*1995*	1989	1965	1973	*1989*	1978	1984
7	1992	1975	1975	1965	*1996*	*1992*	1972	*1996*	1974
8	1983	1974	*1991*	1978	*1990*	1991	*1996*	1981	1978
9	1978	1971	1977	*1992*	1989	1972	1983	*1990*	*1991*
10	*1990*	1977	1984	1973	1972	1984	1964	1959	*1992*

Note: 1997 data incomplete for some stations. (See Fig. 5.1 for river locations)

(Wright, 1978). In the light of subsequent droughts, these very depressed runoff rates now appear substantially less outstanding. For some, mostly eastern, rivers the return period associated with the 1976 minima have been halved. Table 5.2, using annual 90-day minima as a yardstick, confirms the prominence of the 1988–97 period in rankings of low flow periods across a representative range of catchments.

In almost all regions, seasonal flow contrasts intensified over the 1988–97 period although, in some permeable lowland catchments, well above average rainfall in some winters maintained healthy groundwater contributions to river flows through the succeeding dry summers. A broad appreciation of the contrasts in flows over the period 1988–97 with those for the preceding record may be gauged from the flow duration curves featured in Figure 5.10. For the Tay, the UK's largest river in discharge terms, the 1988–97 curve is appreciably steeper than that for the preceding record. By contrast, the contemporary duration curve for the Great Stour, a river draining a largely impermeable catchment in Kent, plots appreciably below that for the pre-1988 period throughout the flow range; mean flows for 1988–97 being around 15% below the preceding average.

5.5.2   *Recent flooding in the UK*

The recent mild climatic conditions have resulted in very limited snow accumulations throughout the UK. A continuing increase in temperature may lead to locally increased melt rates but generally implies a lessening flood risk from this source. In some, mostly eastern, regions, the persistence of notable soil moisture deficits well into the autumn has also tended to reduce the period during which the risk of flooding is most severe. On the other hand, exceptionally high rainfall totals in a number of recent winters have

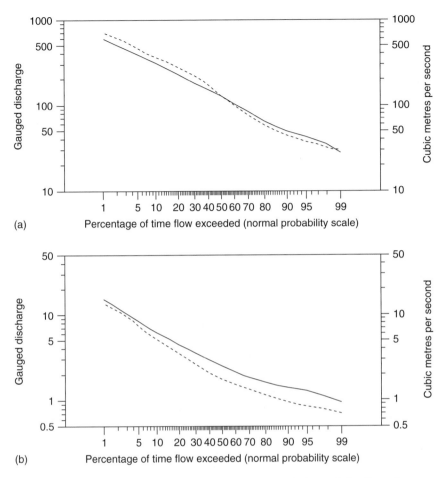

Figure 5.10 Flow duration curves for the (a) Tay at Ballathie and (b) Great Stour at Horton (solid trace: pre-1988; broken trace: 1988–97).

been associated with an increased frequency of spate conditions, in northern Britain especially. The case studies outlined below illustrate how the recent rainfall patterns have generated a sequence of notably damaging flood events.

Six of the twelve wettest winters (December–February) in the Scottish rainfall series, which begins in 1869, now cluster in the 1989–98 period. One consequence is that over the 20 years ending in 1995, parts of Scotland recorded average annual runoff totals up to 30% above the preceding average. A two-fold increase in the number of high flow events over the 1970–94 period was reported for the Rivers Spey, Dee and Annan (Werritty, 1998). The relatively low runoff in the early 1970s is a contributory factor but it is noteworthy that land use change in these catchments has been

Table 5.3 Exceptional flood events in Scottish rivers, 1988–97.

Date	Rivers(s)	Causes, effects and reference(s)
7/2/89	Ness	Widespread heavy frontal rainfall, producing new UK 2-day maximum fall of 306 mm at Kinloch Hourn; new Ness peak flow in intermittent flow record from 1930; Ness rail bridge collapse in Inverness (Inglis, 1989)
1989–90	Spey	Series of high winter maximum flows each causing damages in rural areas (Sprott and McKenna, 1992)
5/2/90	Tay and Earn	Catchment-wide flooding due to heavy 4-day rainfall with wet antecedent conditions + snowmelt: 70-yr return period assessed for Tay at Caputh; >£3 m in damages, >80 homes flooded (Falconer and Anderson, 1993)
1990–93	Teith	Series of winters with three annual maxima assessed with return periods >35 years; general increase in flood frequency since 1988 (Evans *et al.*, 1995)
17/1/93	Tay and Earn	Snowmelt + rainfall event caused by widespread melt of deep snowpack combined with precipitation from two warm fronts; Tay peak flow >2200 $m^3 s^{-1}$; >400 homes flooded; damages estimated at £30 m (Black and Anderson, 1994)
11/12/94	Clyde	Regional flood due to continuous 48-hour rainfall, affecting all main rivers and smaller watercourses across Strathclyde area; 700 homes flooded; total damages estimated at £100 m (Black and Bennett, 1995)
2/7/97	Lossie and Isla	Summer frontal rainfall affecting all north-draining rivers between Nairn and Isla; flow return periods of up to 55 years; damages >£20 m (Brown and Black, 1997)

minimal. Such increases in flood frequency together with the substantial property damage associated with several recent high magnitude events in Scotland has generated considerable public concern.

Table 5.3 lists the rivers and events in which recent flooding has been the subject of particular attention. Three of the five largest rivers in Scotland (the Tay, Ness and Clyde), have achieved new maximum recorded flows since 1988, each having in excess of 40 years of flow records. Also, it should be noted that all the events in Table 5.3, except the July 1997 Lossie/Isla flood, occurred on rivers draining from the western part of Scotland. The cluster of recent exceptionally high annual maximum flows on the Tay has substantially reduced the return period associated with the higher magnitude floods (Figure 5.11), an example of the hydrological shift associated with increased precipitation over western Scotland. By contrast, major peaks in the large eastern Tweed and Dee catchments have been conspicuous by their absence.

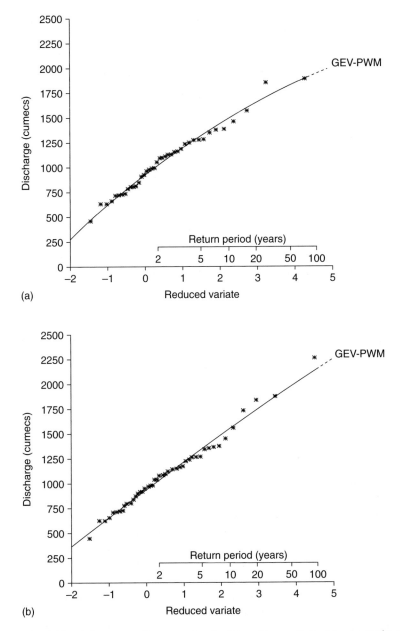

Figure 5.11 Flood frequency diagrams for the River Tay at Ballathie: (a) 1948–87; (b) 1948–97. (The analysis was undertaken using the Institute of Hydrology's HYFAP packge – using a General Extreme Value distribution).

The level of damages arising from the floods featured in Table 5.3 has been heavily controlled by the presence of urban areas, and so the single most important event in this respect has been the December 1994 flood event affecting many parts of the Glasgow conurbation. Key aspects of the damage – insurance claims exceeded £100 million – included the inundation of 700 homes by numerous watercourses across the area (sometimes due to urban drainage problems), two fatalities at a collapsed road bridge, flooding of industrial premises and rail transport disruption in Glasgow for eight months (Black and Bennett, 1995). In floods such as the Tay/Earn event of 1993, more restricted urban floodplain development ensured more limited losses (though no less pernicious at North Muirton in Perth), and the agricultural impact assumed greater importance with 52 km^2 of farmland inundated (Black and Anderson, 1994).

Prior to the exceptionally damaging flooding in the south Midlands over Easter 1998, when levels on the Warwickshire Avon were the highest in a series from 1848, relatively few extreme events had occurred in England over the preceding decade. In many permeable catchments, this quiescence has been linked with the notable low groundwater levels which has been a feature of much of this period. However, on occasions watertables have risen to near-record levels. An exceptionally wet autumn and early winter in 1993 generated very steep seasonal groundwater recoveries in southern England. By late December water-tables were approaching the surface over wide areas and some high level springs flowed for first time in more than 30 years. Continuing heavy rainfall in early 1994 triggered one of the most disruptive and protracted flood events of the recent past in southern England. As a number of wells and boreholes in the Chalk of the South Downs began overflowing, the River Lavant was transformed into a much more rainfall-responsive stream, causing very protracted flooding in Chichester. Although the peak flow in the river reached only 8.1 m^3 s^{-1}, the limited channel capacity of approximately 2 m^3 s^{-1} contributed to the flooding of some 50 properties and the closure of all major roads passing through the area. Had the Chichester city centre culvert carrying the Lavant collapsed, the impact of the event would have increased greatly with the estimated flooding of 1200 properties and evacuation of 10 000 people (Taylor, 1995). The damage and disruption caused by the prolonged nature of the extreme runoff (extending over five weeks) underlines the role of development in increasing the hazard in any floodplain area.

In this section some preliminary assessments of regime variation based on recent hydrological patterns have been presented. The inherent variability of the UK climate implies that the hydrological characteristics of any 10-year period are likely to differ appreciably from the long-term average. The significance of the recent departures is however given greater weight by their broad similarities with a number of favoured climate change scenarios. It is now more difficult to assume that the volatility of the recent past will automatically be superseded by a return to generally more stable conditions.

Nonetheless the sequence of relatively dry winters in parts of Britain since the mid-1990s provides a timely reminder that any trend over short periods must be treated with considerable caution. Only in retrospect will it be possible to confirm that the recent past has provided valuable insights into conditions which may occur with greater frequency in a warmer world.

5.6 A perspective on the present and the future

5.6.1 Environmental sustainability

Climate has always exercised a dominant control over flow regimes and mitigating the effect of too much or too little rainfall will remain a priority task for water management. But as the environmental and ecological value of aquatic systems becomes more fully recognised, and 'sustainability' moves from a theoretical formulation to practical application, so man will necessarily exercise an increasingly active role in managing river systems. In many rivers, the provision of near natural, or perceived 'valuable' biological communities, may be of overriding importance. In others, the focus may be more aesthetic, where, for example, a river provides the mainspring for an urban regeneration programme.

Although the Environment Agency's water resources licensing policies will be a major mechanism in protecting threatened rivers and wetlands, the adoption of low flow remediation measures may be expected to be played out across a broader canvas in the future. Changes in flow regimes (river morphometry also) have a potentially wide range of effects on the ecology of watercourses. Developing means of assessing ecologically acceptable flow regimes is therefore a focus of considerable current research effort. This is, in large part, a response to field evidence. At some sites ecological damage has been found to be very limited, even where relatively large abstraction rates have obtained over many years. In contrast, at other sites where only minor changes in abstraction rates have been recorded, the impact on the ecology has been significant, reflecting the sensitivity of certain species to changes in the hydrological regime.

A range of survey techniques is available for assessing instream ecology but none of the methodologies is truly predictive in terms of specific impacts of flow reductions or changes to channel morphology. The PHABSIM model, (Elliott, *et al.*, 1998) specifically considers ecological demands when appraising flow regimes or recommending remedial options. The system simulates the relationship between streamflow and available physical habitat for specific target species, commonly salmonid fish. Physical habitat will vary between rivers and will also depend on the differing habitat requirements of the target species (and their various life stages). Due to the currently high costs of applying the technique, use of PHABSIM has been restricted to seriously affected river reaches or where licence applications for major abstractions are under consideration.

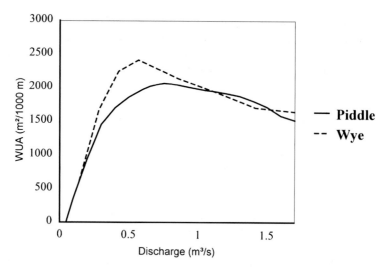

Figure 5.12 Instream physical habitat available for fry/juvenile brown trout in the Rivers Piddle and Wye.

An example of the interactions between river discharge and instream habitat is presented in Figure 5.12. The graph shows the changes in instream physical habitat for the fry/juvenile life stage of brown trout in two contrasting UK rivers, the Piddle, a chalk stream in Dorset, and the Wye, an upland river in an impermeable catchment in Powys, Wales. In both cases the relationship between physical habitat (indexed by weighted useable area, WUA) and discharge is non-linear with habitat availability peaking at approximately $0.7 \, \text{m}^3 \, \text{s}^{-1}$ in the River Piddle and at approximately $0.6 \, \text{m}^3 \, \text{s}^{-1}$ in the Wye. At discharges above and below these levels, physical habitat availability declines. In much of the river, flow depths and velocities are reduced under low discharge conditions to levels below those required by the target species/life stage, and at high discharges velocities become too high for the fish to hold station and they must seek refuge in marginal areas. However, river flow alone, although often perceived to be the cause of ecological stress, may be only one of several factors which in combination impact on the availability of suitable habitat. Differences in water quality, the degree of channelisation and other physical characteristics of streams may interact with regime changes in a complex way. Detailed study of the River Piddle for instance, revealed that sedimentation changes – triggered, in part, by cattle watering practices – was a significant influence on the decline in the river's character and in the fish population it could sustain.

RIVPACS, a multivariate statistical model, has also been applied to the analysis of perceived degradation arising from river regulation, either by impoundment or abstraction (Petts *et al.*, 1991; Armitage and Petts, 1992). Based on sampled invertebrate data (identified to family level), it can calculate the environment quality of a site by comparison to a broad database of over 600 near-unimpacted reference sites. Results of applying the technique to low flow impacted sites have so far been mixed. The technique is able to demonstrate impacts in extreme cases, but is currently not sensitive enough to detect more moderate artificial hydrological alteration. Future development or more ecologically appropriate hydrological indices will undoubtedly enable teasing out of the effects of regulation from those of background natural water quality variations and pollution.

As flow characteristics change, secondary attributes of the flow, such as water depths and velocities, and substrate characteristics may change also. This may lead to changes in species composition, or it may alter biomass, or life history strategies. In extreme cases, this may mean that certain species can no longer be found within a particular river. Figure 5.13 shows flow and physical habitat duration curves (illustrating the time that given levels of flow and physical habitat are equalled or exceeded) for fry/juvenile brown trout on the River Allen, a Dorset chalk stream that historically has suffered from reduced river flows arising from over-abstraction of groundwater. These curves are presented for the predicted habitat availability under actual (i.e. with flow levels artificially reduced) and naturalised conditions (i.e. with the artificial influence on flow removed, in this case by using a groundwater flow model). They demonstrate how physical habitat is reduced for the fry/juvenile brown trout under summer low flow conditions in particular. Due to the non-linear relationship between physical habitat and flow, the habitat reduction is far greater than the flow reduction. This may be of particular relevance to a target species if it produces a 'habitat bottleneck' where the lack of habitat for a particular life stage at a critical time of year means that the overall population of that species is limited.

5.6.2   River flow objectives: a framework for the future

To make sound decisions about the allocation of runoff to potentially competing uses, a set of management objectives for watercourses or river systems needs to be established. A framework for development and implementation of objectives has been established by Acreman and Adams (1998). This may be considered at two levels. First, the overall management objectives must be set. Second, the river flows required to meet the overall objectives need to be defined. For key habitats, the overall management objectives may be governed by the conservation targets of the UK Biodiversity Action Plan or by designations that provide legislative protection, such as Special Areas of Conservation, Special Protection Areas or Sites of Special Scientific Interest (SSSIs). Overall management objectives for individual sites may be established through public or user participation, such as that undertaken during the development of Local Environment Agency Plans (LEAP). The LEAP identifies issues, and often solutions, within a catchment. This is undertaken

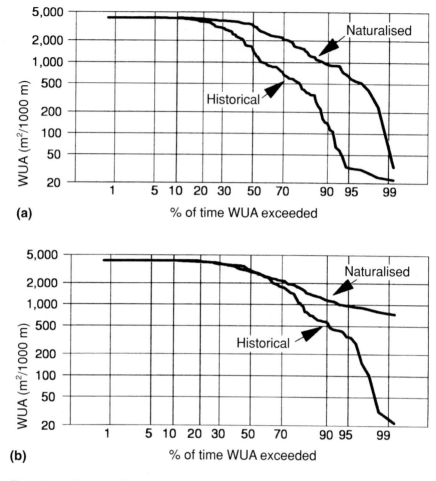

Figure 5.13 Example flow and habitat duration curves for fry/juvenile brown trout in the River Allen under artificially influenced (historical) and naturalised (simulated) conditions: (a) summer months (April–September); (b) winter months (October–March).

in collaboration with the stakeholders in each target catchment including water companies. Objectives are then typically based on these issues. They may be related to amenity, ecology, water supply, navigation, fisheries, aesthetic appeal or archaeology and heritage status (Box 5.1). The LEAP approach marks a move away from the dominant influence exercised by particular sectional interests on particular rivers or reaches in the past. In its place is a more holistic framework which recognises the importance of public perception and social choice, and, commonly, the requirement to arrange and manage public access to important communal assets.

Box 5.1 **Examples of objectives for rivers** (after Acreman and Adams, 1998)

Fisheries	Healthy self-reproducing brown trout population
Plants	Vegetation assemblages
Invertebrates	Invertebrate assemblages
Amenity	Canoeing, boating, bathing, walking
Navigation	Lockage
	Adequate depth
Waste disposal	Effluent dilution
Water supply	Low cost
	High quality
	Reliability
	Protection of existing licences

Notwithstanding the range of potentially relevant considerations, a primary objective, especially at designated sites (e.g. SSSIs), is often to conserve particular species or assemblages in target streams or reaches. Species conservation may also be an important general objective throughout a river system. Although the index species often chosen – the brown trout – may, or may not, be native to the stream or reach under consideration, the objective of many chalk streams has traditionally been to maintain a healthy fishery, notably the presence of catchable large adult fish. This results partly from the use of the health of the fishery as an indicator of overall environmental health, (in many catchments it also reflects the economic importance of fishing to the riparian land owners and political power of the fishing lobby). The river flow objective is then set which will maintain the fishery. For rivers where the management objectives are broader, such as to maintain a healthy aquatic ecosystem, setting appropriate river flow requirements is more complex as different species may have different needs.

In terms of practical application, river reaches under review can be evaluated by numerical scoring of the relevant individual uses. Such an approach is used in the Surface Water Abstraction Licensing Practice (SWALP) being developed by the Environment Agency. Whatever the overall objectives, the river flow requirements needed to meet the objectives must be defined. As river flows are reduced, the ability of the river to meet the overall objectives, whether for dilution, navigation or to support wildlife, also declines. An early form of river flow objective was the Minimum Acceptable Flow (MAF), introduced in the 1963 Water Act. Subsequent amending legislation (Water Resources Act 1989, Environment Act, 1995) has not significantly altered the MAF concept, although the duties of the regulatory authority have been refined (Petts, 1996). The MAF was originally developed as a tool for water

---

Box 5.2 Methods of setting objectives (after Dunbar *et al.*, 1997, Petts *et al.*, 1995)

Tier	Method	Example	Scale
0	Look-up table	Objective commonly related to discharge, indexed to seasonal flows (e.g. during fish migration)	Broad-scale national or regional assessment
1	Desk-top hydrological analysis	E.g. Naturalisation of historical flows and its relationship to existing biological information (such as invertebrate or fish surveys)	↑
2	Multi-disciplinary expert panel	Expert opinion, often combined with additional hydrological and biological data collection and analysis	↓
3	Detailed biological response modelling	E.g. physical habitat modelling assessment	Detailed site or reach

---

resources management, primarily the protection of the rights of downstream water users (including abstraction and navigation). No statutory MAFs have ever been set, but some use is made of non-statutory prescribed flows. Where a minimum flow requirement has been set, it has invariably been determined by hydrological, rather than ecological, criteria. Since many rivers experience low flows, or even dry up naturally during droughts, a minimum flow is of limited utility unless it is set within a probabilistic framework.

Any effective management strategy needs to take account of the natural variability in flows in a river. In the Anglian Region of the Environment Agency, the concept of hands-off flows has been developed, i.e. below a certain flow, abstractions are restricted so that the flow regime is broadly natural or acceptable (i.e. sufficient for other uses downstream). Petts *et al.* (1996) advocated the application of river flow objectives as a management tool, which could lead to the setting of seasonally varying MAFs. This may involve defining a target flow duration curve for a river.

In their international review of methods of setting river flow objectives, Dunbar *et al.* (1997) built on this approach and identified a four-tier strategy for setting objectives (Box 5.2). The type of procedure to follow in particular circumstances will depend on the scale and detail of results required and agreement amongst the stakeholders. In practice, tier 0 would be appropri-

Box 5.3 **Examples of objectives and approaches used**
(after Acreman and Adams, 1998)

	Overall management objective	*Flow/level objective*	*Approach used*
River Babingley	To maintain a wild brown trout population Water supply	Ecologically acceptable flow duration curve	Physical habitat modelling (PHABSIM) and naturalised flow duration curve from rainfall-runoff model
River Kennet	To maintain a wild brown trout population Water supply	Flow should not fall below that which results in a reduction in physical habitat for brown trout of more than 10%	Physical habitat modelling (PHABSIM)
River Avon	Protect salmon migration Water supply	Minimum flows at critical times of the year	Radiotracking of salmon

ate for national or regional monitoring and identifying rivers where further study would be required. However, the use of such a general methodology is unlikely to be acceptable to all stakeholders where water resources are under pressure. In such cases, the defined flow will need to be set using a more robust technique – tier 3 – (e.g. using a the PHABSIM model), particularly if it is to stand up to a public inquiry. It should be emphasised that models such as PHABSIM are incremental and do not specify what level of change is significant or a threshold below which a target species or ecosystem cannot be sustained. Consequently, the actual level of an acceptable flow or target regime characteristics remains a political or social decision (Jacobs, 1997). The need to adapt procedures to the circumstances encountered will be evident from Box 5.3 which gives examples of the objectives set and the alleviation options chosen, for individual rivers and wetlands. Proposed enhancements to the nature conservation potential of a watercourse need to ensure that compensating flood defence action is not required to counteract any significantly increased upstream flood risk. Through time, river enhancement strategies will also need to accommodate any climate-driven trends in flow patterns or water quality characteristics; greater seasonality in flow patterns and increased water temperatures would, for example, impact particularly on game fisheries.

5.7   Conclusion

Man has always been dependent on rivers and at the mercy of their capriciousness. As the number of pristine river systems declines further, artificial influences on flow regimes increase and development pressures on floodplains become ever more compelling, so the degree to which rivers depend on man is brought into sharper focus. In order to exercise a responsible stewardship over UK river systems, sophisticated strategies will be required to reconcile the multiplicity of demands on them with the need to protect and improve their ecological status and aesthetic appeal. This challenge is heightened by the potential impacts of climate change. In Scotland, over the past 10 years it has become clear that patterns of flooding are by no means constant and several extended drought episodes in the English lowlands have served as a reminder of our vulnerability to relatively modest reductions in rainfall. Fortunately our understanding of catchment and river behaviour is increasing, helping to establish the knowledge base to develop more environmentally sympathetic management options. For 'sustainability' to achieve real practical expression it is important to improve the UK's capability to distinguish between anthropogenic and climate-driven regime changes, and reinforce a catchment planning framework that is able to exploit hydrological science to the full.

References

Acreman, M.C. (1989) Extreme rainfall in Calderdale, *Weather*, 44, 438–44.

Acreman, M.C. and Adams, B. (1998) Lowflows, groundwater and wetland interactions. Report to the Environment Agency (W6-013) UKWIR(98/WR/09/1) and NERC (BGS WD/98/11).

Anon. (1995) Long Flow Records, in *1994 Yearbook, Hydrological Data UK Series*. Institute of Hydrology, Wallingford. pp. 35–36.

Anon. (1996) *1995 Yearbook, Hydrological Data UK Series*. Institute of Hydrology, Wallingford.

Archer, D.R. (1992) *Land of Singing Waters: Rivers and Great Floods of Northumbria*. Spredden Press, Stocksfield. pp. 138–41.

Armitage, P.D. and Petts, G.E. (1992) Biotic score and prediction to assess the effects of water abstraction on river macroinvertebrates for conservation purposes. *Aquatic Conserv.: Mar. Freshwater Ecosyst.*, 2, 1–17.

Arnell, N.W. and Reynard, N.S. (1996) The effect of climate change due to global warming on river flows in Great Britain. *J. Hydrol.*, 132, 321–42.

Barber, K.E. and Coope, G.R. (1987) in Gregory, K.J., Lewin, J. and Thornes, J.B. (eds), *Palaeohydrology in Practice*. Wiley, Chichester. pp. 217–50.

Barber, K.E. and Twigger, S.N. (1987) Late Quaternary Palaeoecology of the Severn Basin, in Gregory, K.J., Lewin, J. and Thornes, J.B. (eds), *Palaeohydrology in Practice*. Wiley, Chichester. pp. 217–50.

Bayliss, A.C. and Jones, R.C. (1993) Peaks-over-threshold flood database: summary statistics and seasonality. *Institute of Hydrology Report No. 121*, Institute of Hydrology, Wallingford.

Biodiversity Challenge Group (1996) *High and Dry: the Impacts of Over-abstraction of Water on Wildlife*.

Black, A.R. (1996) Major flooding and increased flood frequency in Scotland since 1988. *Phys. Chem. Earth*, 20, 463–68.

Black, A.R. and Anderson, J.L. (1994) The great Tay flood of January 1993, in *1993 Yearbook, Hydrological Data UK Series*. Institute of Hydrology, Wallingford. pp. 29–34.

Black, A.R. and Bennett, A.M. (1995) Regional flooding in Strathclyde, December 1994, in *1994 Yearbook, Hydrological Data UK Series*. Institute of Hydrology, Wallingford. pp. 29–34.

Black, A.R. and Werritty, A. (1997) Seasonality of flooding: a case study of North Britain, *J. Hydrol.*, 195, 1–25.

Brandon, J. (ed.) (1996) *Palaeohydrology: Context, Components and Application*. British Hydrological Society Occasional Paper No. 7. British Hydrological Society, Wallingford.

Brown, R.C. and Black, A.R. (1997) Moray Floods, 1–3 July 1997. *Circulation*, 55, 1–2.

Bye, P. and Horner, M. (1998) Easter 1998 Floods. Report by the Independent Review Team to the Board of the Environment Agency.

Climate Change Impacts Review Group (CCIRG) (1996) *Review of the Potential Effects of Climate Change in the United Kingdom, Second Report*. HMSO, London.

Crooks, S.M. (1994) Changing flood peaks on the River Thames. *Proc. Inst. Civ. Eng., Water, Maritime Energy*, 106, 267–97.

Curran, J.C. and Robertson, M. (1991) Water quality implications of an observed trend of rainfall and runoff. *J. IWEM*, 5, 419–25.

Dobbie, C.H. and Wolf, P.O. (1953) The Lynmouth flood of August 1952. *Proc. Inst. Civ. Eng.*, 522–88.

Dunbar, M.J., Gustard, A., Acreman, M.C. and Elliott, C.R.N. (1997). Overseas approaches to setting river flow objectives. Report to the Environment Agency. (W6-161), and NERC. Institute of Hydrology, Wallingford.

Dury, G.H. (1977). Underfit streams: retrospect, perspect and prospect, in Gregory, K.J. (ed.), *River Channel Changes*. Wiley, Chichester. pp. 280–93.

Elliott, C.R.N., Dunbar, M.J., Gowing, I.M. and Acreman, M.C. (1998) A habitat assessment approach to the management of groundwater dominated rivers. *Hydrological Processes*, 13, 12–21.

Evans, T.E., Smith, T.M. and Gill, B. (1995) Flood management of upper Teith basin, above Callander, in *Proceedings of the British Hydrological Society 5th National Hydrology Symposium*. British Hydrological Society, Wallingford. pp. 3.11–19.

Falconer, R.H. and Anderson, J.L. (1993) Assessment of the February 1990 flooding in the River Tay and subsequent implementation of a flood-warning system. *J. IWEM*, 7, 134–48.

Foster, M.J., Werritty, A. and Smith, K. (1997) The nature, causes and impacts of recent hydroclimatic variability in Scotland and Northern Ireland, in *Proceedings of the British Hydrological Society 5th National Hydrology Symposium*. British Hydrological Society, Wallingford. pp. 8.9–17.

Green, F.H.W. (1971) History repeats itself – flooding in Moray in August 1970. *Scottish Geog. Mag.*, 87, 150–52.

Green, S. and Marsh, T.J. (1997) A consideration of rainfall, runoff and losses at Plynlimon in the context of long term hydrological variability in the UK and maritime Western Europe, *Hydrol. Earth Syst. Sci.*, 1(3), 399–407.

Green, S., Sanderson, F.J. and Marsh, T.J. (1996) Evidence for recent instability in rainfall and runoff patterns in the Celtic regions of Western Europe, in *Hydrologie dans les pays celtiques*, INRA (Les Colloques No. 79), Paris, 73–83.

Gregory, K.J., Lewin, J. and Thornes, J.B. (1987) *Palaeohydrology in Practice*. Wiley, Chichester.

Gustard, A., Bullock, A. and Dixon, J.M. (1992) Estimating Low River Flows in the United Kingdom. *Institute of Hydrology Report No. 108*. Institute of Hydrology, Wallingford.

Higgs, G. (1987) Environmental Change and Hydrological Response: Flooding in the Upper Severn Catchment, in Gregory, K.J. *et al.* (eds), *Palaeohydrology in Practice*. pp. 131–160.

Hollis, G.E. (1975) The effect of urbanisation on floods of different recurrence intervals. *Water Resources Res.*, 11, 431–35.

Inglis, T. (1989) River flows and flood warning, in *Proceedings of the East Highland Floods Sympsium*, Dingwall, October 1989. Paper A3.

Institute of Hydrology (1999) *Flood Estimation Handbook*. Institute of Hydrology, 5 vols.

Jacobs, M. (1997) *Making Sense of Environmental Capacity*. Council for the Protection of Rural England. London.

Jones, P.D., Olgilvie, A.E.J. and Wigley, T.M.L. (1984) *River Flow Data for the United Kingdom: Reconstructed Data back to 1844 and Historical Data back to 1556*. Climatic Research Unit. University of East Anglia, Norwich.

Kirby, C., Newson, M.D. and Gilman, K. (1991) Plynlimon research: the first two decades. *Institute of Hydrology Report No. 109*. Institute of Hydrology, Wallingford.

Lamb, H.H. (1977) *Climate: Present, Past and Future*, Vol. 2: *Climate History and the Future*. Methuen, London.

Lewin, J. (1981) Contemporary Erosion and Sedimentation, in Lewin, J. (ed.), *British Rivers*. Allen and Unwin, London.

Littlewood, I.G. and Marsh, T.J. (1996) A re-assessment of the monthly naturalised flow record for the River Thames at Kingston from 1883, and implications for the relative severity of historic droughts, *Regulated Rivers: Res. Manag.*, 12, 13–26.

Marsh, T.J. (1996a) River Flow and Groundwater Level Records – the Instrumented Era, in *Palaeohydrology: context, components and application*. British Hydroloical Sociey Occasional Paper No. 7. British Hydrological Society, Wallingford. pp. 7–24.

Marsh, T.J. (1996b) The 1995 Drought – evidence of climatic instability? *Proc. Inst. Civ. Eng. Water, Maritime Energy*, 118, 189–95.

Marsh, T.J. and Lees, M.L. (eds) (1998) *Hydrometric Register and Statistics 1991–95*. Hydrological Data UK Series. Institute of Hydrology, Wallingford.

Marsh, T.J., Monkhouse, R.A., Arnell, N.W., Lees, M.L. and Reynard, N.S. (1994) *The 1988–92 Drought*. Hydrological Data UK Series. Institute of Hydrology, Wallingford.

Mayes, J.C. (1995) Changes in the distribution of annual rainfall in the British Isles. *J. CIWEM*, 9, 531–39.

Newson, M.D. (1975) *Flooding and Flood Hazard in the United Kingdom*. Oxford University Press, Oxford. p. 11.

NERC (1990). *Water Quality in the Environment*. Natural Environment Research Council.

National Rivers Authority (1993) *Low Flows and Water Resources*. National Rivers Authority, Bristol.

Petts, G.E. (1996) *Linking Hydrology and Ecology*: River Wissey Investigations. Report to the National Rivers Authority, Bristol.

Petts, G.E., Armitage, P.D., Forrow, D., Bickerton, M., Castella, E., Gunn, R. and Blackburn, J.H. (1991) The effects of abstractions from rivers on benthic invertebrates, CSD Report no. 1230, Nature Conservancy Council, Peterborough.

Petts, G.E., Crawford, C., Clarke, R. (1996) Determination of Minimum Flows. Environment Agency R&D Note 449. Environmental Research and Management, University of Birmingham.

Raven, P.J., Fox., Everard, Holmes, N.T.H. and Dawson, F.H. (1996) River Habitat Survey: a new system to classify rivers according to their physical character, in Boon, P.J., *Freshwater Quality: Defining the Indefinable*. HMSO, Stirling.

Robinson, (1990) Impact of improved land drainage on river flows. *Institute of Hydrology Report No. 113*. Institute of Hydrology, Wallingford.

Robinson, M. (1998) 30 years of forest hydrology changes at Coalburn: water balance and extreme flows. *Hydrol. Earth Syst. Sci.*, 2(2) 233–38.

Robson, A.J. and Reed, D.W.R. (1996) Non-stationarity in UK flood records. Flood Estimation Handbook Note 25. Report to Ministry of Agriculture, Food and Fisheries, London.

Robson, A.J., Jones, T.K. Reed, D.W. and Bayliss, A.C. (1998) A study of national trend and variation in UK floods. *Int J. Climatol.*, 18, 165–82.

Sprott, W.C. and McKenna, E. (1992) River Spey flooding in 1989 and 1990 and subsequent recommendations. Paper presented at joint IWEM/SHG/SHSG meeting, Perth, 31 March 1992.

Taylor, S.M. (1995) The Chichester Flood, January 1994. *1994 Yearbook, Hydrological Data UK Series*, Institute of Hydrology, Wallingford. pp. 23–27.

Wright, C. (1978) Synthesis of river flows from weather data. *Central Water Planning Unit Technical Note No. 26*.

Werritty, A. (1998) Hydroclimatic variability in Scotland – trends, potential causes and impacts. Scottish Hydrological Group Meeting. *Circulation*, 57, 5–6.

6

RIVER WATER QUALITY

Richard Williams, Tim Burt and Geoff Brighty

This chapter reviews changes in water quality by reference to three river water contaminants each of which cover problems that have been evident over a range of time scales. Nutrient runoff has been of concern for tens of years, pesticides which have risen to prominence over the last decade and endocrine disrupters, the impact of which have only recently come to light.

6.1   Introduction

The natural water quality of a river will be determined primarily by the catchment soil type and underlying geology to which water, falling on the catchment as rain, is exposed as it drains to the river. Deviations from this baseline water quality are generally caused by the influence of people through point and diffuse pollution sources. Diffuse sources include increased sulphur and nitrogen in atmospheric deposition and nitrate, phosphorus and pesticides from agricultural land. Sewage treatment works discharging nitrate, nitrite, ammonia, BOD, phosphorus, surfactants and steroid oestrogens and industrial effluents containing micro-organics, heavy metals and solvents are the point sources of main concern.

The river basin research element of the Land–Ocean Interaction Study (LOIS; Wilkinson *et al.*, 1997) has identified the main influences on water quality for the rivers draining into the North Sea by analysis of Environment Agency data (e.g. Robson and Neal, 1997a; Robson and Neal, 1997b) and data from a purpose designed sampling network (Leeks *et al.*, 1997 (for the network); e.g. Neal *et al.*, 1997; House *et al.*, 1997 (for the analysis). The patterns of water quality described were attributable to the influences described above and can be considered representative of catchments of mixed land use – urban, industrial and agricultural – typical of the UK.

6.2   Nutrients

It is now widely acknowledged that agriculture is the main source of river and groundwater pollution in the rural areas of lowland UK. This has been known for some time as far as nitrate pollution is concerned (Royal Society, 1983; Burt et al., 1993), but there is now also widespread concern about eroded soil and associated pollutants such as phosphorus and pesticides (see Chapter 2).

The House of Lords' report *Nitrate in Water* (1989) began by commenting on the conflicts which can arise when the use of land for farming comes into conflict with the use of land for water supply. Concern initially focused on alleged links between high nitrate concentrations in drinking water and two health problems in humans: the 'blue-baby' syndrome *methaemoglobinaemia* and gastric cancer. More recently, problems associated with nutrient enrichment of fresh and marine waters have been recognised and attention has turned to phosphate as well as nitrate.

The popular misconception that the nitrate problem is caused by farmers applying too much nitrate fertiliser is too simplistic. Nevertheless, there is now little doubt that the high concentrations of nitrate in fresh waters noted in recent years have mainly resulted from runoff from agricultural land and that the progressive intensification of agricultural practices, with increasing reliance on the use of nitrogenous fertiliser, has contributed significantly to this problem. Since 1945, agriculture in the UK has become much more intensive: fields are ploughed more frequently; more land is devoted to arable crops, most of which demand large amounts of fertiliser; grassland too receives large applications of fertiliser and stocking densities are higher; many low-lying fields are now underdrained, encouraging more productive use of the land and speeding the transport of leached nitrate to surface water courses. It is true that lowland rivers receive larger quantities of nitrogen from sewage effluent, and southeast England is more affected by atmospheric deposition of nitrogen; even so, budgeting studies confirm that agriculture is the main source of nitrate in river water (Burt and Johnes, 1997).

Betton et al. (1991) have mapped nitrate concentrations for mainland Britain. A marked northwest to southeast gradient is evident, reflecting relief, climatic conditions and agricultural activity. Upland areas in the north and west are usually characterised by nitrate concentrations below $1 \text{ mg NO}_3\text{-N l}^{-1}$. This reflects the high rainfall and low temperatures of such areas. Soils tend to conserve organic matter and mineralisation rates are low. In contrast, a decreasing ratio of runoff to rainfall and an increasing intensity of agricultural land use towards the south and east of Britain results in higher mean concentrations of nitrate in river water. Many of the lowland rivers are characterised by concentrations above $5 \text{ mg NO}_3\text{-N l}^{-1}$; in East Anglia and parts of the Thames basin, mean nitrate concentrations in rivers are close to the EC limit of $11.3 \text{ mg NO}_3\text{-N l}^{-1}$, a level exceeded in some spring waters especially in the Jurassic limestones of the Cotswolds and Lincolnshire Wolds (Johnes and Burt, 1993).

The changing pattern of lowland agriculture since 1945 is reflected in long-term records of nitrate for surface and ground waters (Johnes and Burt, 1993). For both large and small rivers, there has been a relatively steady upward trend in nitrate concentrations, often of the order of 0.1–0.2 mg NO_3-N l^{-1} a^{-1}. Analyses for relatively short time series of just a few years (e.g. Betton et al., 1991) have shown that the upward trend may be interrupted, either because of climatic variability (drier years are associated with lower nitrate concentrations) or because of land use change. Nevertheless, statistical analysis of long time series shows that the main effect is a steady increase in nitrate levels over time which is independent of climate (Johnes and Burt, 1993). If trends continue, the mean nitrate concentration of many rivers in the UK will soon be above the EC limit. In many cases, this level is already exceeded during the winter when nitrate concentrations reach their maximum. In catchments where groundwater is the dominant discharge source, this long-term trend may be prolonged since it may take years for nitrate to percolate down to the saturated zone. In such basins, nitrate pollution may remain a problem for decades to come. In recent years, a number of options have been considered as a means of halting the upward trend. Catchment-wide schemes encourage farmers to undertake good farming practice. More localised schemes, like the Nitrate Sensitive Area scheme and the Nitrate Vulnerable Zones, involve greater restrictions on farming practice and financial compensation may be available to farmers for loss of income (Burt et al., 1993). Much interest currently focuses on the use of riparian land as nitrate buffer zones (Haycock et al., 1997; Box 9.2).

Rather less attention has been devoted to phosphorus pollution and there is greater uncertainty about the relative contributions of point and non-point sources. Phosphorus in soil may reach rivers in a number of forms. Much may be adsorbed to stream sediments, whereas a smaller but significant fraction may be present in dissolved, bioavailable form, especially where conditions on the stream bed or in riparian soils favour mobilisation of phosphorus incorporated in previously stable forms (Goulding et al., 1996). Working in small rural catchments, Johnes et al. (1996) showed that livestock wastes were the largest contributors to both nitrogen and phosphorus exports, with cultivation the next most important source of nitrogen and people of phosphorus; this suggested that small sewage treatment works were not important sources of phosphorus. They argued that sewage effluent is more likely to be the predominant source of phosphorus in large rivers. However, Foster et al. (1996) showed that diffuse sources of sediment-associated phosphorus still represented a significant proportion of the *total* phosphorus load transported in the Warwickshire Avon, especially during times of high flow. Nevertheless, though soils can provide significant amounts, sewage effluent remains the main source of *dissolved* phosphorus (Foster et al., 1996). As with nitrate, there is currently much interest in the ability of riparian zones to buffer phosphorus losses from farmland. Unfortunately, it may be that conditions which favour denitrification (saturated, anoxic soils) are not conducive to retention of phosphorus and further research is required on the ability of riparian buffer zones to sustain accumulation of phosphorus over the long term.

6.3   Pesticides

The late 1970s saw the development of annual grass weed herbicides that allowed a general shift to winter cereal production systems. Such systems carry less risk for farmers because preparation of seedbeds is much easier in the drier conditions prevailing in late summer/early autumn. However, herbicide applications to winter cereals usually occur around late October and through November when the ground is near field capacity and subsequent winter rains are likely to lead to pesticide leaching or runoff. The pollution of UK surface waters with pesticides resulting from their use as part of normal agricultural production was demonstrated in a number of independent field studies carried out the late 1980s and early 1990s (Williams *et al.*, 1995; Johnson *et al.*, 1994; Harris *et al.*, 1993). As expected, the main pollutants were herbicides with those applied in autumn generally giving rise to higher concentrations than those applied in spring.

It is probably true to say that one of the biggest factors in driving forward studies of pesticide fate and behaviour was the European Union's Drinking Water Directive (Council of the European Communities, 1980). This Directive stipulates a very strict Maximum Admissible Concentration (MAC) for any single pesticide in potable waters of 0.1 µg l^{-1} and 0.5 µg l^{-1} for all pesticides. It was evident from the studies mentioned above that concentrations in excess of these values were common in small streams draining agricultural land and there was concern for concentrations in larger rivers used for water supply. In samples collected from controlled waters, around 100 pesticides were detected. However, the frequency of detection of individual pesticides above the MAC rarely exceeded 15% with the vast majority below 5%. Table 6.1. lists the ten pesticides that exceeded the MAC most often in 1994 and 1995 (Department of the Environment, Transport and the Regions, 1997).

Methods of reducing diffuse pollution from agricultural land by management practice are receiving increased attention (Harris, 1995). These include: (1) lower pesticide usage through altering the area treated by changing cropping patterns; (2) integrated crop management in which cultivation techniques for pest reduction (e.g. crop rotation) are used in conjunction with more targeted pesticide applications, all aimed at reducing usage; (3) altering tillage methods to produce seed beds that are less prone to runoff or leaching and (4) creating buffer strips at field boundaries adjacent to streams or drainage ditches to intercept runoff.

The majority of pesticide runoff research has been carried out on plots and fields and there are many measurements available of pesticide loads at this scale. The processes that lead to these loads are well understood although they have proved too complex to model accurately. The conversion of field runoff values to concentrations likely to be observed in surface waters in large catchments is currently missing. This is an important omission if the

Table 6.1 Top 10 pesticides most frequently exceeding 0.1 µg/l in controlled waters in 1994 and 1995 (DETR, 1997).

1994[a]			*1995*		
Pesticide[c]	*Total number of samples*	*Per cent samples >0.1 µg/l*	*Pesticide*[c]	*Total number of samples*	*Per cent samples >0.1 µg/l*
Isoproturon	3374	18	Isoproturon	3188	19
Mecoprop	2121	16	Diuron	2573	14
Diuron	2426	13	Mecoprop	2924	10
2,4-D[b]	603	8	2,4-D[b]	1637	6
Chlorotoluron	2387	6	Chlorotoluron	2613	5
Simazine	4324	5	Simazine	6005	4
Dicamba	281	4	Penta-chlorophenol	5349	3
Atrazine	4387	4	Atrazine	6203	3
Penta-chlorophenol	6553	4	Bentazone	1057	2
Lindane	8280	1	Lindane	7237	2

Notes: [a] Bentazone had a 14% exceedance, but the number of samples was very low (64) and probably the result of a special survey. [b] This pesticide is approved for use in or near water at a rate that would result in a concentration above 0.1 µg/l. [c] All the pesticides listed are herbicides with the exception of lindane which is an insecticide and pentachlorophenol which has insecticidal, fungicidal and herbicidal activity.

intervention measures described above are to be assessed in terms of say protecting surface water drinking supplies. Some attempts are being made to address this through GIS tools, e.g. the Environment Agency's Prediction of Pesticide Pollution in the Environment (POPPIE) model, and the Severn Trent Water Catchment Information System model (CatchIS) (Hollis *et al.*, 1995). However, both these systems share a semi-empirical approach to their predictions of pesticide runoff relating them to generalised pesticide usage collated at the parish scale. Therefore, there is no explicit routing of pesticides from field boundaries through the water course network. This approach is pragmatic and sensible given the lack of knowledge of the fate and behaviour of pesticides in small ditches and river systems and how these might ameliorate their concentrations. There is a need for the processes occurring in these environments to be understood and their effects incorporated into improved estimates of pesticide concentrations throughout the water course network of large catchments.

Sheep husbandry is another source of pesticide pollution of surface waters with both the siting of sheep dipping sites and the disposal of the spent dip giving problems. Impacts of organophosphorus-based sheep dips on stream fauna have been reported (e.g. Virtue and Clayton, 1997). Because of the additional problem of the implication of organophosphorus insecticides in human health effects, operators are now required to apply for a licence of competence before they can be used. Uptake has been slow and new, pyrethroid-based sheep dips have come on the market for which no such licence is required. This presents problems from a stream pollution perspective as these chemicals are toxic to stream fauna at very low concentrations. An example of this toxicity has been reported in the Tweed catchment where a small spillage of dilute cypermethrin waste dip was accidentally discharged to the Slitrigg Water killing 1200 fish and severely damaging invertebrate populations for 5 km downstream (Virtue and Clayton, 1997).

Pesticides are also used non-agriculturally, usually for weed control on hard surfaces (e.g. railway lines, car parks, motorway verges). These uses of pesticides have also been implicated in surface water pollution and have resulted in concentrations above the EC MAC in water works in London (White and Pinkstone, 1993). The persistent herbicides atrazine and simazine were the main products in use for this purpose until their banning in 1993. In anticipation of and since the ban, they have been replaced by diuron which is now increasing in surface waters as the other chemicals decrease. There is some evidence that ground work by water companies and chemical manufacturers may have discouraged such widespread use of diuron as was the case with atrazine (White and Pinkstone, 1995; NRA, 1995). In this case, the control of the non-agricultural use of pesticides has only served to highlight an increasing trend in isoproturon concentrations associated with agricultural use (Figure 6.1). In the Thames region as a whole, atrazine and simazine accounted for 77% of drinking water limit contraventions while in 1994 isoproturon alone accounted for 68% (White and Pinkstone, 1995).

Pesticides for both agricultural and non-agricultural uses are continually being developed and new products are being registered for use every year. The trend is for pesticides to be applied at much lower doses while still maintaining the same efficacy. While this might reduce the number of exceedances of the EC MAC, the increased activity of the compound may mean they have a detrimental effect in the environment even at low levels. It is therefore essential that research is continued to understand the levels at which these new and existing chemicals impact on stream fauna and that methods are developed for prediction of probable exposure levels.

6.4   Endocrine disrupters

Animals and plants have an internal system of chemical messengers (hormones) and receptors that control their basic functioning, termed the *endocrine* system. It is responsible for a wide range of physiological processes, such as growth, metabolism, homeostasis and reproduction. In animals, this system consists of a number of glands that secrete the hormones that are then transported, usually in the bloodstream to their target organ(s) (Baulieu and Kelly, 1990). At the target organ, the hormone molecule exerts its effect

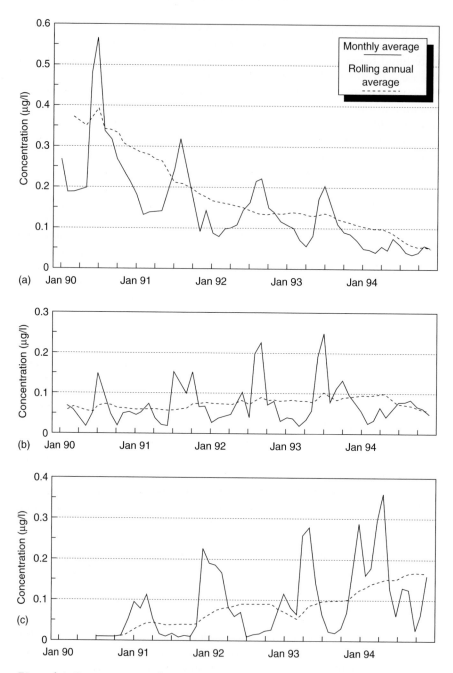

Figure 6.1 Concentrations of (a) atrazine, (b) diuron, and (c) isoproturon in drinking water supplies from a water treatment works serving London. (From
White and Pinkstone (1995).)

by binding to its receptor and initiate physiological change, e.g. to produce a protein molecule, stimulate cell division or alter metabolic rate.

The endocrine system shows a conserved evolution, e.g. some steroid hormones present in some invertebrate groups (e.g. molluscs) are also found in vertebrates, such as fish, amphibia and mammals producing similar effects in each organism. However, there are also differences between the endocrine systems of animal groups, reflecting the diverging evolutionary pathways and the organism's adaptive physiology (Le Blanc, 1998). A simple example would be the hormone systems controlling exoskeleton moulting in invertebrates, known collectively as *ecdysones*, which are not found in vertebrates.

Whilst the functioning of the endocrine system is broadly understood, it is also known that certain chemicals can specifically interfere with the endocrine system, either by mimicking the action of hormones (Nimrod and Benson, 1994) or by interfering in some other way. These chemicals are collectively referred to as *endocrine-disrupting substances* (European Commission, 1996). They are believed to affect the health, growth and reproduction of a wide range of organisms by interfering with the normal functioning of the endocrine (hormonal) system (Institute for Environment and Health, 1995).

Endocrine disruptors can affect the endocrine system in several ways. Some substances bind directly to a hormone receptor, to elicit a response. Such substances therefore act as 'false hormones', e.g. nonylphenol mimics the action of oestrogen hormones (Routledge and Sumpter, 1996). Others bind to receptors but do not lead to a response, thereby blocking the receptor to the true hormone producing an antagonistic response, such as DDE on androgen receptors (Kelce *et al.*, 1995). Other substances, e.g. the antifouling paint active ingredient tributyltin (TBT), interfere with the hormone synthesising enzymes (TBT arrests aromatase activity), and lead to an imbalance of the steroid hormones (Matthiessen and Gibbs, 1998). At this stage, how an organism's health is affected by endocrine disrupting substances is not fully understood, but due to the complexity of the hormone system and the increasing number of substances with such properties, the potential risks for ecological damage are high.

In the UK, evidence for endocrine disruption focuses largely on the aquatic environment, in particular rivers that receive sewage effluent. Roach (*Rutilus rutilus*) caught in the settlement lagoons of a sewage treatment works (STW) and the River Lea, London, and analysed as part of a routine survey, were found to have an 'abnormally' high incidence of a 'hermaphrodite' (intersex) condition, that is, oocytes were found within the testes (Thames Water, 1981). This indicated that the males had been 'feminised', a response characteristic of exposure to female-type hormones, the oestrogens. Reference sites did not reveal any abnormalities, and a causal link was hypothesised between inputs to the rivers, particularly sewage effluents, and oestrogenic substances that they may potentially contain.

Later work investigating the oestrogenic effects of sewage effluent has shown that both male trout (*Oncorhynchus mykiss*) and carp (*Cyprinus carpio*) produce the egg yolk protein vitellogenin when exposed to the effluent for 21 days (Purdom *et al.*, 1994). The protein is normally produced only by mature female fish in response to oestrogen within their bloodstream, and its presence is a very sensitive and specific biomarker for exposure to oestrogens (Sumpter and Jobling, 1995), although this is not a permanent marker as the intersex condition. Virtually all sewage effluents showed some oestrogenic activity, suggesting that it was a common factor that was causing the response.

When sewage treatment works effluents had been confirmed as inducing oestrogenic responses in fish, concern then focused on the impacts in the receiving rivers. Caged test fish placed at varying distances downstream exhibited similar effects as seen in undiluted effluent, such effects as elevated blood vitellogenin levels. However, in some locations, attenuated testis growth rate, reduced testis size and increased liver size were also observed (Harries *et al.*, 1996). At other sites, even though the effluent was oestrogenic, no effects were observed in the test fish caged in the river, presumably because the active chemicals in the effluent were either degraded rapidly, or diluted below a concentration that affected the fish. The oestrogenic substances were not identified in most of these field studies, but at one site, where a 5 km stretch was strongly oestrogenic to fish, the response was attributed to very high levels of nonylphenol (and related chemicals) emanating from local textile processors (Harries *et al.*, 1995). Local agreements secured by the then National Rivers Authority (now Environment Agency) have since reduced nonylphenol inputs and the oestrogenic activity within the river. Overall, the conclusions of this research were that industrial effluents may play a dominant role in certain situations, but that domestic wastewater input into the STW effluent was a common factor amongst all effluents assessed and was a likely consistent source of oestrogenic activity on receiving waters.

Identifying the causative agents for such impacts in complex mixtures is a demanding exercise. Either suspected agents can be analysed for (with the risk that the causal substance(s) is missed), or one can screen the mixture against a test assay in order to target the biological activity and the chemical nature of the substance. Although many substances have been shown to have oestrogenic properties, what was being released to the environment was unclear. An investigation into the identity of oestrogenic substances in domestic STW effluent was undertaken using a fractionation approach, allied to an *in vitro* recombinant yeast oestrogen screen (Desbrow *et al.*, 1998). This involved producing individual fractions of sewage effluent, testing these in the assay that would respond to oestrogenic substances. Using this approach a single fraction was identified as containing the biologically-active substances. These were three steroid hormones; the *natural* steroids 17β-oestradiol and oestrone, were detected on all sampling occasions within a range from 1 to 75 ng l^{-1}, and ethinyl oestradiol, a *synthetic* hormone used in contraceptive and hormone replacement therapy treatments, was found on a third of the occasions and at much lower concentrations, approximately one-tenth of that of the other steroids.

Further tests confirmed that, at these concentrations, induction of vitellogenesis in male trout and roach does occur (Routledge *et al.*, 1998). There was some evidence of additivity of response by the fish to mixtures of natural steroids. The source of the natural steroids was believed to be human, and the presence of free (unconjugated) hormone indicated a biotransformation from the excreted bound (conjugated) form in the sewer and/or during treatment. The research confirmed the major role of steroid hormones in the oestrogenic activity of wastewaters and raised the potential for a widespread influence on the reproductive health of fish. However, the exposure of fish to oestrogens in the environment, as demonstrated by vitellogenin production, does not necessarily indicate a causal link to other physiological effects, such as the intersex condition.

On-going research has surveyed the incidence and spatial extent of the intersex condition in roach (Jobling *et al.*, 1998). Results have confirmed that the intersex condition is widespread in English rivers compared with still waters and most river control sites. At some river sites, all male roach sampled were intersex. A mean value for males below STW effluent discharges showing this effect was 61%, compared with a mean for males at upstream sites of 28%. Control sites had less than 18% incidence, and were as low as 4%, with a mean value of 12%. Plasma vitellogenin levels were also elevated in the intersex fish compared with that of 'normal' males, suggesting that vitellogenin was being produced endogenously by the intersex fish. Gonad weight was lower in fish from sites downstream of STW discharges.

The severity of intersex abnormalities in males also differed between sites. For example, the number of eggs within the testis varied, some male fish having a few primary eggs whereas others had many eggs and in some cases also had oviducts, the reproductive tract normally found in females. These results were based on a scoring analysis termed the *intersex index*, where a normal male was counted as zero, with intersex scores increasing with increasing numbers, size and stage of oocyte development and the presence of a female reproductive tract, and full female scoring as 7.

Fish from the majority of sites had a low intersex index value, with very few oocytes within their testes and normal male ducts. Thus suggests, but does not prove, that reproductive function would not be impaired. However, the highest severity and incidence of effect were at sites that received high quantities of effluent, or where the effluent received the least dilution. Fish from these sites showed gross effects on the testis such that there is doubt over their functionality. These results constitute the first report of widespread reproductive abnormalities in fresh waters in the UK.

There was a strong positive correlation ($p<0.0002$, $r^2 = 0.68$) between the proportion of sewage effluent in the river and the number of male fish

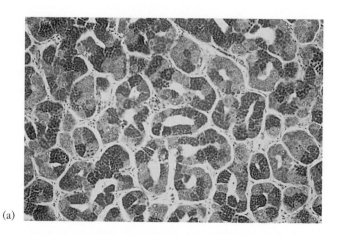

(a)

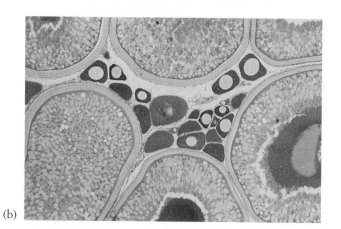

(b)

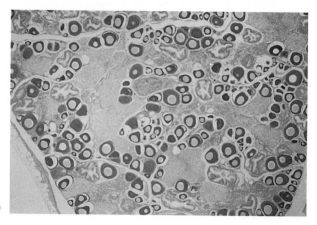

(c)

showing signs of the intersex condition. This confirms the role of effluents as a major causal factor in the evolution of the intersex condition in fish. A statistically significant but less strong correlation ($p < 0.0001$; $r^2 = 0.312$) was determined between the proportion of sewage effluent in the river and the severity of effect in the males. Taken together, these data indicate that where effluents receive limited dilution after discharge, male fish are at high risk of being exposed to oestrogenic effluents and subsequently becoming intersex.

The association of severity of effect is expected to be weaker since it measures the degree of response and relies more on quantifying the exposure profile. Improving this correlation requires dose response information for the causal substances, and knowledge of the timing and duration of exposure. By correlating against data for fish of the same age at each site, the timing and duration of exposure could be accounted for as all fish would have had a similar life history. However, to achieve this level of resolution in a mixed age population, large numbers of fish are needed.

The spatial extent of effects shown by this study is of concern, given that most rivers receive effluent and thus could be affected. Whilst focusing on one species, other limited assessments within the programme indicate that more cyprinid species are also being affected. The significance of the findings at the population level are not known, and research is continuing into the reproductive health of the fish. However, there is a strong potential for adverse reproductive effects through impaired spermatogenesis and resultant sperm quality (quantity and motility).

The present state of information suggests that sexual disruption is widespread within UK rivers. Clearly such responses could be indicative of risks to population health. Whilst this may not be a threat to more common fish species, those fish already under threat from other pressures, and those with high conservation value, must be targeted to ensure that populations are sustainable.

Placing this issue in context within an overall improving water quality base in UK is not straightforward since although the fish are not apparently killed by exposure to oestrogenic endocrine disrupting substances, the effects observed appear to be permanent and of relevance to the future of the population. It is important, however, still to focus on those water quality stresses that lead

---

Plate 6.1 (*opposite*) (a) Normal male testis: a transverse section through a male roach testis. Note regular form of sperm-producing tissue (spermatogonia) and lumens that lead to sperm duct (Geoff Brighty). (b) Normal female ovary: primary (dark-stained cytoplasm, small) oocytes and secondary (larger, yolky) oocytes with their light-pink stained nucleii. The secondary oocytes will be released at the next spawning (Geoff Brighty). (c) Intersex male testis: a tranverse section through an apparently 'normal' (from external examination) male testis. Within, areas of normal male tissue (light pink) are interspersed with large numbers of primary oocytes, to greater than 50% of the section area (Geoff Brighty).

to mortality before trying to address non-lethal impacts such as endocrine disruption. A balanced approach is therefore required to understand the pressures for the aquatic environment and set realistic priorities for addressing them.

6.5   Conclusions

There is no doubt that the work of hydrologists in collaboration with chemists and agricultural scientists has improved enormously levels of understanding of the mechanisms of pollution of surface water in recent times. However, there are still areas where further elucidation of process or application of knowledge is required:

- in establishing the relative importance of sewage treatment works and agricultural inputs of phosphorus to river;
- in improving knowledge of the functioning of riparian buffer strips and an assessment of their ability to protect water courses from nitrate, phosphorus and pesticide runoff;
- in developing methods for the reliable estimation of pesticide concentrations at the catchment scale;
- in improved assessments of the impacts (both acute and chronic) of pollutants on aquatic flora and fauna, particularly with respect to bioavailabilty of sediment bound micro-organics;
- in assessing the impacts of endocrine disrupting chemicals on fish populations as opposed to the established effects on individuals.

Water quality issues are likely to become more important as demand for water and the concomitant increase in waste water becomes greater. This is likely to be further exacerbated by climate change impacts giving rise to longer, low-flow conditions in UK summers. Nutrient inputs may stimulate greater biological activity under these conditions and micro-organic inputs will be diluted into lower river flows thus increasing concentrations and possibly both chronic and acute effects. Sub-lethal, long-term effects of micro-organic contaminants are likely to become increasing of concern. In addition, because these compounds are generally hydrophobic, the role of sediments in controlling exposure levels to stream fauna is likely to become an important topic.

References

Baulieu, E-E. and Kelly, P.A. (eds) (1990) *Hormones, from Molecules to Disease*. Chapman and Hall, London.

Betton, C., Webb, B.W. and Walling, D.E. (1991) Recent trends in NO_3-N concentration and load in British rivers. *IAHS Publication 203*, 169–80.

Burt, T.P. and Johnes, P.J. (1997) Managing water quality in agricultural catchments. *Trans. Inst. Br. Geog.*, NS 22(1), 61–68.

Burt, T.P., Heathwaite, A.L. and Trudgill, S.T. (1993) *Nitrate: Processes, Patterns and Management*. Wiley, Chichester.

Council of the European Communities (1980) Directive of the 15 July 1980 relating to the quality of water intended for human consumption. 80/778/EEC; 05 1 229, 30 August 1980.

Department of the Environment, Transport and the Regions (1997) *Digest of Environmental Statistics, No. 19*. HMSO, London.

Desbrow, C., Routledge, E.J., Brighty, G.C., Sumpter, J.P. and Waldock, M. (1998) Identification of oestrogenic chemicals in STW effluent. 1. Chemical fractionation and in vitro biological screening. *Environ. Sci. Technol.*, **32**, 1549–558.

European Commission (1996) *European Workshop on the Impact of Endocrine Disrupting on Human Health and Wildlife*. Report of proceedings of workshop, Weybridge. EUR 17549.

Foster, I.D.L., Baban, S.M.J., Wade, S.D., Buckland, P.J. and Wagstaff, K. (1996) Sediment-associated phosphorus transport in the Warwickshire Avon. *IAHS Publication 236*, 303–12.

Goulding, K.W.T., Matchett, L.S., Heckrath, G., Webster, C.P., Brookes, P.C. and Burt, T.P. (1996) Nitrogen and phosphorus flows from agricultural hillslopes, in Anderson, M.G. and Brooks, S.M. (eds), *Advances in Hillslope Processes, Vol. 1*. Wiley, Chichester. pp. 213–28.

Harries, J.E., Jobling, S., Matthiessen, P., Sheahan, D.A. and Sumpter, J.P. (1995) *Effects of Trace Organics on Fish – Phase 2*. Report to the Department of the Environment, Foundation for Water Research, Marlow. Report no. FR/D 0022.

Harries, J.E., Sheahan, D.A., Jobling, S., *et al.* (1996) A survey of estrogenic activity in United Kingdom inland waters. *Environ. Toxicol. Chem.*, **15**, 1993–2002.

Harris, G.L. (1995) Pesticide loss to water – a review of possible agricultural management opportunities to minimise pesticide movement, in Walker, A., Allen, R., Bailey, S.W. *et al.* (eds), *Pesticide Movement to Water. BCPC Monograph No. 62*. British Crop Protection Council, Farnham. pp. 371–80.

Harris, G.L., Bailey, S.W., Rose, S.C., Mason, D.J. and Llewellyn, N. (1993) The transport of pesticide residues to surface waters in small clay-based catchments, in *Proceedings of the Brighton Crop Protection Conference – Weeds 1993*. British Crop Protection Council, Farnham. pp. 815–20.

Haycock, N.E., Burt, T.P., Goulding, K.W.T. and Pinay, G. (1997) *Buffer Zones: Their Processes and Potential in Water Protection*. Quest Environmental, Harpenden.

Hollis, J.M., Keay, C.A., Hallett, S.H. and Gibbons, J.W. (1995) Using CatchIS to assess the risk to water resources from diffusely applied pesticide, in Walker, A., Allen, R., Bailey, S.W. *et al.* (eds), *Pesticide Movement to Water. British Crop Protection Council Monograph No. 62*. British Crop Protection Council, Farnham. pp. 345–50.

House of Lords (1989) *Nitrate in Water: Sixteenth Report from the European Communities Committee*. HMSO, London.

House, W.A., Leach, D., Long, J.L.A. *et al.* (1997) Micro-organic compounds in the Humber rivers. *Sci. Total Environ.*, **194/195**, 357–72.

Institute for Environment and Health (1995) *Environmental Oestrogens: Consequences to Human Health and Wildlife. Assessment A1*. Institute for Environment and Health, Leicester.

Jobling, S.J., Nolan, M., Tyler, C.R., Brighty, G. and Sumpter, J.P. (1998) Widespread sexual disruption in wild fish. *Environ. Sci. Technol.*, **32**, 2498–506.

Johnes, P.J. and Burt, T.P. (1993) Nitrate in surface waters, in Burt, T.P., Heathwaite, A.L. and Trudgill, S.T. (eds), *Nitrate: Processes, Patterns and Management*. Wiley, Chichester. pp. 269–317.

Johnes, P.J., Moss, B. and Phillips, G. (1996) The determination of total nitrogen and total phosphorus concentrations in freshwaters from land use, stock headage and population data: testing of a model for use in conservation and water quality management. *Freshwater Biol.*, 36, 451–73.

Johnson, A.C., Haria, A.H., Bhardwaj, C.L. *et al.* (1994) Water movement and isoproturon behaviour in a drained heavy clay soil: 1. persistence and transport. *J. Hydrol.*, 163, 217–31.

Kelce, W.R., Stone, C.R., Laws, S.C. *et al.* (1995) Persistent DDT metabolite *p,p* = -DDE is a potent androgen receptor antagonist. *Nature*, 375, 581–85.

Le Blanc, G.A. (1998) Steroid hormone-regulated processes in invertebrates and their susceptibility to environmental endocrine disruption, in Guillette, L.J. (ed.), *Environmental Endocrine Disruptors: an Evolutionary Perspective*. Taylor and Francis, London.

Leeks, G.J.L., Neal, C., Jarvie, H.P., Casey, H. and Leach, D.V. (1997) The LOIS river monitoring network: strategy and implementation. *Sci. Total Environ.*, 194/195, 101–10.

Matthiessen, P. and Gibbs, P.E. (1998) Critical appraisal of the evidence for tributyltin-mediated endocrine disruption in mollusks. *Environ. Toxicol. Chem.*, 17, 37–43.

Neal, C., Robson, A.J., Harrow, M. *et al.* (1997) Major, minor, trace element and suspended sediment variations in the River Tweed: results from the LOIS core monitoring programme. *Sci. Total Environ.*, 194/195, 173–92.

Nimrod, A.C. and Benson, W.H. (1994) Environmental estrogenic effects of alkylphenol ethoxylates. *Crit. Rev. Toxicol.*, 26, 335–64.

NRA (1995) Pesticides in the Aquatic Environment. *National Rivers Authority Water Quality Series Report No. 26*. HMSO, London.

Purdom, C.E., Hardiman P.A., Bye, V.J. *et al.* (1994) Estrogenic effects of effluents from sewage treatment works. *Chem. Ecol.*, 8, 275–85.

Robson, A.J. and Neal, C. (1997a) A summary of regional water quality for Eastern UK rivers. *Sci. Total Environ.*, 194/5, 15–37.

Robson, A.J. and Neal, C. (1997b) Regional water quality of the river Tweed. *Sci. Total Environ.*, 194/5, 173–92.

Routledge, E.J. and Sumpter, J.P. (1996) Estrogenic activity of surfactants and some of their degradation products assessed using a recombinant yeast screen. *Environ. Toxicol. Chem.*, 15, 241–49.

Routledge, E.J., Sheahan, D., Desbrow, C., Brighty, G.C., Sumpter, J.P. and Waldock, M. (1998) Identification of oestrogenic chemicals in STW effluent. 2. In vivo responses in trout and roach. *Environ. Sci. Technol.*, 32, 1559–65.

Royal Society (1983) *The Nitrogen Cycle of the United Kingdom*. Royal Society, London.

Sumpter, J.P. and Jobling, S. (1995) Vitellogenesis as a biomarker for estrogenic contamination of the aquatic environment. *Environ. Health Perspect.*, 103 (Suppl. 7), 172–8.

Thames Water Authority (1981) Hermaphrodite roach in the River Lea. Thames Water, Lea Division.

Virtue, W.A. and Clayton, J.W. (1997) Sheep dip chemicals and water pollution, *Sci. Total Environ.* **194/5**, 207–17.

White, S.L. and Pinkstone, D.C. (1993) Amenity and industrial use of herbicides: the impacts on drinking water quality, in *Proceedings of the Brighton Crop Protection Conference – Weeds 1993*. British Crop Protection Council, Farnham. pp. 363–68.

White, S.L. and Pinkstone, D.C. (1995) The occurrence of pesticides in drinking water, in Walker, A., Allen, R., Bailey, S.W. *et al.* (eds), *Pesticide Movement to Water: BCPC Monograph No. 62*. British Crop Protection Council, Farnham. pp. 263–8.

Wilkinson, W.B., Leeks, G.J.L., Morris, A. and Walling, D.E. (1997) River and coastal research in the Land–Ocean Interaction Study. *Sci. Total Environ.*, **194/195**, 5–14.

Williams, R.J., Brooke, D.N., Matthiessen, P. *et al.* (1995) Pesticide transport to surface waters within an agricultural catchment. *J. Inst. Water Environ. Manag.*, **9**, 72–81.

7

GROUNDWATER

Brian Adams, Ian Gale, Paul Younger, David Lerner and John Chilton

This chapter is about that part of the hydrologic cycle which occurs below the ground surface, i.e. groundwater. It discusses the management of groundwater storage, the pollution of groundwater by nitrate and pesticides from agriculture, the acidic and metalliferous waters that flow from abandoned mine workings and the issues associated with urban development and industrial change, including rising groundwater in urban areas from reductions in pumping.

7.1   Introduction

In the north and the west of the UK, the relatively impermeable ancient (Pre-Cambrian and Palaeozoic) rocks are less conducive to groundwater movement and storage than the more permeable, younger (Mesozoic and Palaeogene) strata to the south and east of England (Figure 7.1). The geology also has an impact on the climate (and hence the potential recharge to groundwater) with the generally older, harder rocks of the northwest having a higher relief and hence generally higher rainfall than the younger softer rocks of the southeast. Thus the upland regions of Britain are generally suitable for surface water development with opportunities for surface water capture in the high runoff areas having high relief. However, even in these regions, groundwater resources can be locally important and are often extensively used for private domestic water supplies. Groundwater provides some 32% of public water supplies in England and Wales, 14% in Northern Ireland and 5% in Scotland. This wide regional variation in use of groundwater for public supply in England and Wales is shown in Figure 7.2; the principal control on this variation being the geology, see Figure 7.1. The importance of groundwater has been recognised for a long time and records of variations in groundwater levels at a number of places have been kept for many years. The longest continuous record is for the Chilgrove House well in the Chalk of southern England where records date back to 1836, part of which is shown in

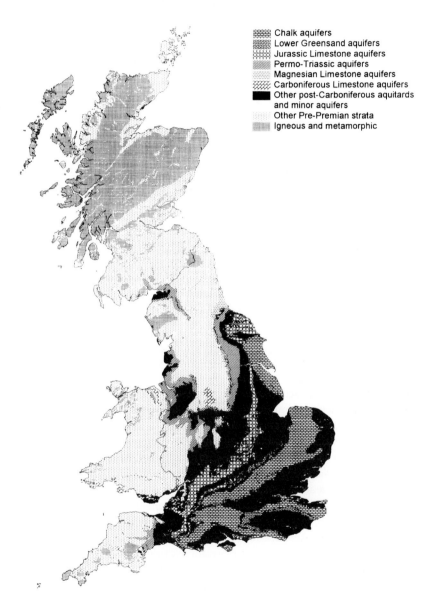

Chalk aquifers
Lower Greensand aquifers
Jurassic Limestone aquifers
Permo-Triassic aquifers
Magnesian Limestone aquifers
Carboniferous Limestone aquifers
Other post-Carboniferous aquitards
and minor aquifers
Other Pre-Premian strata
Igneous and metamorphic

Figure 7.1 Principal geological formations in UK.

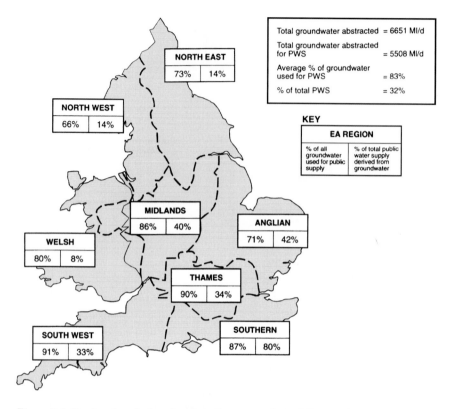

Total groundwater abstracted = 6651 Ml/d

Total groundwater abstracted for PWS = 5508 Ml/d

Average % of groundwater used for PWS = 83%

% of total PWS = 32%

KEY

EA REGION

% of all groundwater used for public supply	% of total public water supply derived from groundwater

NORTH EAST — 73% | 14%

NORTH WEST — 66% | 14%

MIDLANDS — 86% | 40%

ANGLIAN — 71% | 42%

WELSH — 80% | 8%

THAMES — 90% | 34%

SOUTH WEST — 91% | 33%

SOUTHERN — 87% | 80%

Figure 7.2 Regional variations in groundwater use for public supply in England and Wales (1997). Modified from NRA (1995). Data source: Digest of Environmental Statistics, DETR (1997). © NERC. All rights reserved.

Figure 7.3. The variation in water level reflects the balance between the recharge (and hence rainfall) and aquifer outflow in southern England.

Whilst groundwater provides around a third of public water supply in England and Wales, it forms less than 15% of the total water abstracted for all purposes. This difference in use reflects the better quality of groundwater compared to surface water. However, whilst the baseline natural quality of groundwater is generally high, there are ever increasing instances of man's activities polluting this valuable resource. In general terms, prior to the mid-1970s groundwater research in the UK was directed towards quantitative aspects of groundwater resources. However, since that time the qualitative aspects have become ever more important. Whilst this is to some extent due to our relatively high level of knowledge of the quantitative aspects of the UK's groundwater resources, it more importantly reflects our increasing realisation of the impact of our activities upon the quality of groundwater.

The modern approach to the development of groundwater in the UK began with the Water Act of 1945. Downing (1993) identifies three phases

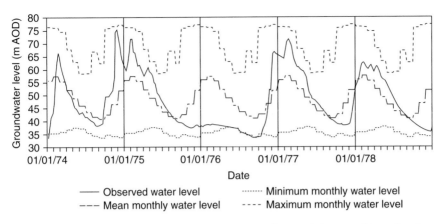

Figure 7.3 Groundwater hydrograph in Chalk aquifer at Chilgrove House showing minimal groundwater recharge in 1975/6. The aquifer was subsequently recharged to above average levels in the winter of 1976/7.

of groundwater development since that Act which relate to the implementation of appropriate legislation:

- an era of resource assessment, 1945–1963;
- an era of groundwater management, 1963–1974;
- an era of groundwater quality, 1974 onwards.

Evidently, these eras identify the main focus of groundwater resources development over a particular time period rather than being totally exclusive of other activities. Thus, groundwater management issues have continued to be important, but arguably not dominant, since 1974. For example, the management of groundwater storage remains an important consideration for water companies and the Environment Agency, and discussion of this topic provides the opportunity for a general introduction to the role of groundwater in the hydrologic cycle (section 7.2).

A major change in the focus of groundwater research issues was heralded by the work of Foster *et al.* (1986) which first highlighted the potential problem of nitrate pollution, although the potential scale of the problem and its implications to the farming community were perhaps not fully appreciated at that time. However, nitrate is not the only agrochemical to have affected groundwater quality and the widespread use of pesticides is now recognised as having a significant impact on our groundwater resources. In section 7.3 the occurrence of nitrate and pesticides in groundwater is discussed and the legislative response (in the case of nitrate) and areas of research described. The growing awareness of these and other types of anthropogenic pollution resulted in the development of a variety of regional groundwater protection polices by individual regional water authorities prior to privatisation of the Water Industry in 1991. Following the establishment of the National Rivers Authority, a National Groundwater Protection Policy was published and is now implemented by the Environment Agency (Chapter 10).

Another major change in groundwater hydrology has resulted from the impact of mining activities on the hydrogeologic regime. The initial dewatering and subsequent flooding of workings upon mine closure both have major implications for groundwater and are discussed in section 7.4.

However, not all the changes in groundwater hydrology are to do with quality. Many of Britain's cities overlie aquifers. Urbanisation usually brings pollution problems (Rivett *et al.*, 1990; Burston *et al.*, 1993; Ford and Tellam, 1994), but also changes groundwater hydrology. Inputs to and outputs from groundwater are all changed by the presence of the city, with consequent effects on the water table position. Typically, water tables under UK cities fell by tens of metres between the Industrial Revolution and the 1960s, but have rebounded since. The rebound has often been enough to cause geotechnical and flooding problems (Brassington, 1990). Section 7.5 looks at the causes and effects of urbanisation on groundwater. It considers, in turn, the ways that water enters an urban aquifer, the changes in the rates of pumping, and the overall effects.

7.2 Management of groundwater storage

The store of water in groundwater reservoirs, aquifers, is constantly being depleted by natural seepage to wetlands (Chapter 8) and rivers, and forms the baseflow component of river flow (Chapter 5). Baseflow varies with underlying geology. In low permeability catchments, baseflow can be insignificant in comparison with the surface runoff component, resulting in river flows closely following precipitation patterns. In permeable catchments with large groundwater storage capacity, baseflow can provide the majority of streamflow. This results in smoothing of the response of streamflow to precipitation over months or years through changes in groundwater storage and travel time between points of recharge and discharge. This variation in stream flow behaviour is illustrated in Figure 5.3, which shows a comparison of the hydrographs of the Rivers Lambourne (groundwater dominated) and Ock (surface runoff dominated).

Natural discharge from aquifers is augmented by abstraction for public water supply, industrial and agricultural usage. This abstraction can intercept groundwater flow towards natural discharge points and hence reduce baseflow in streams and rivers. However, much of the groundwater taken for public supply and some industrial uses is returned to the surface water system as treated waste water. Careful management of the timing and location of the abstraction and the location of discharge of return waters can minimise the impacts.

Groundwater is frequently taken near to the centre for demand, often near rivers, and is collected and returned to rivers as treated wastewater many kilometres down gradient, and hence down catchment. In coastal areas, discharge is usually directly to the sea. In extensive water supply/wastewater collection systems, large quantities of groundwater, as well as surface water, can be taken from one catchment and discharged to another, reducing the flow in both the upper and lower reaches of the stream. Increased urban development in the upper reaches of catchments will tend to exacerbate the problem further.

Natural replenishment of groundwater storage is largely from infiltration of precipitation through permeable geological formations to recharge the underlying aquifers. Except during some storm events, recharge can generally only occur when evapotranspiration and the soil moisture deficit are satisfied. In the UK, significant recharge of groundwater only occurs during the winter months from October to March but the period and quantity of recharge can vary considerably, both spatially and temporally. Some recharge also occurs from leakage through the beds of streams where they cross permeable rocks and the water table is below the stream bed.

The impacts of climate change on groundwater recharge, and subsequent baseflow to streams, is therefore a major concern. The current predictions of warmer, drier summers and warmer, wetter winters (Chapter 1) with more variability and extreme events could have a variety of impacts on groundwater storage. The warmer, wetter winters could provide a net increase in recharge but the warmer, drier summers could place more demand on groundwater resources and delay the onset of recharge. The greater variability and increase in extreme events could result in more frequent winters with little recharge as have been experienced recently. Quantifying the storage capacity of aquifers is key to managing water supply through periods of drought.

Other anthropogenic impacts on groundwater recharge include land use change (Chapter 2), irrigation and urbanisation. Changes to deeper rooted crops with higher water demand can reduce infiltration, and hence recharge, through interception and increased transpiration. Drainage of lowland and wetlands can increase aquifer depletion with a corresponding increase in baseflow. Irrigation will result in greater infiltration but this water can contain elevated concentrations of nutrients and pesticides.

Afforestation of lowland catchments will impact recharge to a greater or lesser extent depending on the type of tree planted, e.g. conifers use significantly more water than ash annually. Replacing grassland with ash could lead to a slight increase in recharge where as conifers would reduce recharge. At Thetford in East Anglia, the forest cover reduced recharge by 50% (Calder, 1992).

Urbanisation of permeable catchments has a complex set of impacts on recharge and runoff. Reduction of permeability through paving and roof run-off leads to increased rapid storm run-off but also to increased localised

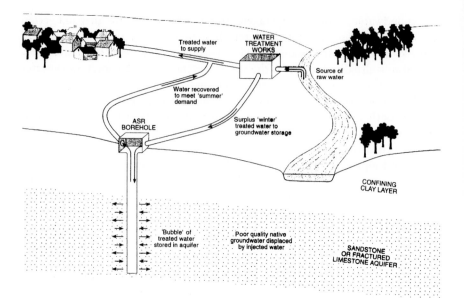

Figure 7.4 Schematic diagram of an ASR scheme.

deep infiltration in soakaways. Leakage from water reticulation, waste disposal and drainage networks can make significant contributions to recharge and have increased the rate of rise of groundwater levels beneath some contributions including Birmingham and Liverpool.

Groundwater/surface water systems have been highly managed for decades. As demands increase for a balance between supply, amenity and environmental needs, our groundwater resources will have to be increasingly managed in innovative ways. Conjunctive use of surface and groundwater can effectively meet demands from variable sources. The Shropshire groundwater scheme, for example, utilises groundwater in the Triassic Sherwood Sandstone to maintain flow in the River Severn during periods of low flow; it is used for supply and environmental benefits. The scheme is only used about one year in three, the groundwater recovering from natural recharge in the intervening periods. The flow of the River Severn is also maintained by releasing water from surface reservoirs, Llyn Clywedog and Lake Vyrawy in Central Wales.

Artificially recharging aquifers has had limited acceptance until recently. Water is induced into aquifers during winter surplus through surface basins or spreading or by injecting in to boreholes (Figure 7.4). The largest scheme to date is the North London Artificial Recharge Scheme which stores treated potable water in the Chalk and Basal Sands aquifers confined beneath London Clay. The scheme comprises thirty six wells and boreholes through which water is both injected and recovered, at 100 Ml d^{-1} during times of shortage.

An extension of this management technique is Aquifer Storage Recovery (ASR) where water is injected, usually into confined aquifers, but these can contain non-potable water. The method works by creating a 'bubble' of treated potable water which is recovered when needed. The schemes have the great advantages that they have minimal environmental impact, can be developed in a staged manner and can resolve a whole range of water supply problems. Trial sites are currently being investigated in several aquifers in the UK.

7.3 The impact of agriculture on groundwater-nitrate and pesticides

7.3.1 Nitrate in groundwater

Nitrate continues to be a groundwater quality management issue in most European countries and in North America, and is increasingly appearing as an issue elsewhere in the world as the monitoring of groundwater quality is developed and extended. In the UK, the extent of nitrate pollution is now well documented. The worst affected areas are the drier eastern and central parts of the country, where the most intensively cultivated areas are also those with lower rainfall and less infiltration to dilute the leached nitrate. These are also the parts of the country in which our major aquifers are located and where the highest proportions of people depend on groundwater for public supply.

Nitrate pollution of groundwater is difficult to control because the sources are diffuse and the transport pathways and processes of transformation are complex. Sources include intensive agricultural use of organic and inorganic nitrogen fertilisers, ploughing up of grassland, disposal of waste from intensive livestock production, urban and industrial wastewater, use of sewage sludge on the land and atmospheric deposition. Of these, it is the nitrate of agricultural origin which provides the principal concern in the UK. Nitrate leached from beneath arable land or intensively managed grassland is transported through the unsaturated zone with the infiltrating recharge to the underlying groundwater. Because subsurface movement is slow, there is often a delay of 10–50 years before water leaving the soil zone reaches groundwater abstraction points. The full impact of increased leaching of nitrate was not felt immediately in supply sources, and there is a corresponding time lag before the beneficial impacts of changes in agricultural practice are felt.

7.3.2 The legislative response

Nitrate has been the focus of important European legislation because elevated nitrate concentrations in water may present problems with respect to both potability and eutrophication. The EC Drinking Water Directive introduced a maximum admissible concentration of nitrate in potable water supplies of 11.3 mg NO_3-N l^{-1} (50 mg NO_3 l^{-1}). After its eventual ratification by the UK Government and implementation into UK legislation, all water supplies intended for human consumption were required to meet this standard. With steadily rising groundwater nitrate concentrations, more and more ground-water sources exceeded the standard; 60 in 1970 rising to 90 in 1980, 105 in 1984, 132 in 1987 (House of Lords, 1989) and 192 in 1990, representing some 1% of the population. Recent reports from the Drinking Water Inspectorate suggest there is 3–4% non-compliance with the nitrate standard, although these data are derived from within the distribution system in 'water supply zones' and cannot be related easily to the actual quality of groundwater in aquifers.

Compliance imposed significant costs on the water industry, as curative measures are expensive to construct and operate. These measures, which include replacement of sources, blending of water from high- and low-nitrate sources, and treatment have been introduced where the EC standard has been exceeded. In the longer term, particularly where nitrate concentrations are still increasing towards the EC standard, there is scope for the introduction of preventive measures to control the amount of nitrate leaching to aquifers. This was recognised by the European Commission in establishing the Nitrate Directive, which established a framework for action to reduce nitrate concentrations in affected areas by designating 'Nitrate Vulnerable Zones' (NVZs) in which surface water or groundwater exceeds or is likely to exceed 11.3 mg NO_3-N l^{-1} if preventive action is not taken. The Directive envisaged that reduction in nitrate concentrations would be brought about by wide-ranging modifications to agricultural practices. Following a long process of public consultation, 65 NVZs were designated around groundwater sources in England and Wales in 1994, and additional such zones are possible as the Directive requires member states to review designations every five years. Action plans to control nitrate leaching are to be developed and implemented by 1999.

The Government established the pilot Nitrate Sensitive Areas (NSAs) in 1990 to test the effectiveness of measures to reduce nitrate leaching. Ten pilot NSAs were selected, representative of the range of farming practices, climatic conditions and hydrogeological situations. Each NSA was defined as the catchment of a groundwater supply source. Within the NSAs, farmers were encouraged to join the scheme, implementing changes in agricultural practice in return for agreed payments. An additional 22 NSAs were designated in 1994.

7.3.3 *Nitrate research*

Research into the occurrence and behaviour of nitrate in the sub-surface has been extensive, both by the agricultural and environmental research

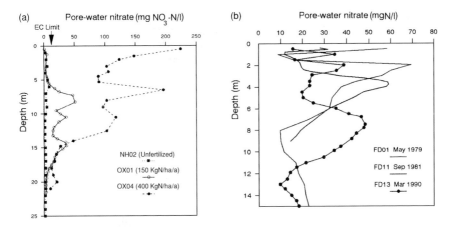

Figure 7.5 Unsaturated zone nitrate profiles in the Chalk aquifer. Example (a) shows increasing nitrate leaching losses beneath more intensively fertilised grassland, and example (b) shows successive profiles beneath cereal cultivation, indicating downward movement of a nitrate peak.

communities. Initially, in the 1970s, efforts were largely concentrated on quantifying nitrate leaching losses. Lysimeters have traditionally been used by agricultural researchers, and many long-term observations of the relationship between land use, nitrogen applications and leaching are based on them (Addiscott *et al.*, 1991). Porous cup suction samplers have also been used. Within the environmental community, investigations have included undisturbed sampling in the unsaturated and saturated zones of the major UK aquifers to observe vertical solute profiles and their evolution with time under a wide range of land uses (Figure 7.5). As a result of all these efforts, a reasonably comprehensive picture of nitrate leaching has been obtained (Burt *et al.*, 1993).

More recently, and in response to the developing legislative framework, research in the agricultural sector has looked at ways in which nitrate leaching losses might be reduced by modifying tillage practices, controlling the amounts and timing of fertiliser applications, avoiding bare ground in autumn and winter and by changing land use to grassland or woodland. Interest has also been focused on the possibility of denitrification as a natural mechanism for reducing the amount of nitrate actually reaching groundwater. Potential denitrifying microbial populations have been isolated from each of the UK's major aquifers, and the likelihood that denitrification occurs in confined aquifers has been demonstrated, but the extent and rate of denitrification under natural conditions in unconfined aquifers is unlikely to be sufficient to have significant beneficial impact on nitrate concentrations.

The water utilities remain concerned about nitrate in groundwater. Where nitrate concentrations are already high, expensive treatment plants to remove nitrate have been installed, because the benefits of land use measures will take too long to come through to the abstraction points. Where nitrate concentrations are moderate, but still rising towards the EC standard, there is considerable scope for control measures at the land surface to help protect the utilities' blending options. They are, therefore, interested in predictions of future groundwater nitrate concentrations, taking account of the range of land-use changes envisaged under the NVZ and NSA measures. There is also interest in the potential impact in terms of nitrate leaching of the disposal to land of sewage sludge, as its disposal at sea will be prohibited by EC legislation from the end of 1998.

7.3.4 *Pesticide occurrence in groundwater*

Advances in the understanding of the processes of nitrate pollution of groundwater naturally led to a consideration of the risk to groundwater quality from other agrochemicals. If nitrate was leaching to groundwater, then it should be possible that, with increasing use, the more mobile pesticide compounds would be leached below the soil and into aquifers. The study of pesticides in the environment does, however, present significant technical difficulties (Foster *et al.*, 1991) because of:

- the wide range of compounds in use, many of which break down to toxic and persistent metabolites, analytical screening for all of which would be prohibitively expensive;
- the need to work at very low concentrations, requiring sophisticated analytical procedures;
- great care is required to avoid sample loss or contamination.

The very stringent maximum admissible concentration of $0.1 \ \mu g \ l^{-1}$ in the EC Drinking Water Directive was effectively a surrogate zero at the time it was proposed in 1980. The problems and costs of devising reliable sampling techniques and analytical methods have meant that routine monitoring data for pesticides in groundwater at concentrations low enough to be related to the Directive have been available only in the last ten years or so.

Early, localised surveys of pesticide occurrence in the mid to late 1980s in the predominantly arable farming areas of eastern England indicated that the most widely used cereal herbicides could be detected in groundwater and surface waters. Mecoprop, MCPA, 2,4-D were amongst the most commonly detected, at concentrations ranging up to $0.2–0.5 \ \mu g \ l^{-1}$. In addition, the triazine herbicide atrazine was widely detected, although it has much more limited agricultural usage. In the 1990s, as monitoring in the public water supply distribution system by the Drinking Water Inspectorate (DWI) and in rivers and groundwaters by the National Rivers Authority and then the Environment Agency has developed, a somewhat clearer picture of the

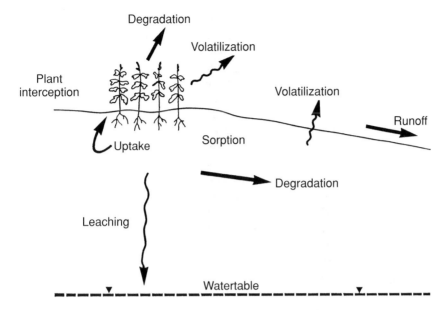

Figure 7.6 Processes determining the fate of pesticides in the soil and unsaturated zone.

occurrence of pesticide residues in the water environment has emerged. This has confirmed that the EC Directive has been exceeded in some public supplies. More than 30 individual compounds have been detected. As for nitrate, the most affected areas are central, eastern and southern England. Concentrations above 1 µg l^{-1} have rarely been recorded.

7.3.5 *Research*

The difficulties of sampling and analysis outlined above in relation to monitoring also apply to research on pesticides in groundwater. Of the processes governing pesticide fate and behaviour shown in Figure 7.6, most are sufficiently active in and above the soil that, for conventional agricultural pesticides, the amount available for leaching is likely to be a very small percentage of that applied. If, however, this small proportion was very persistent and/or reached the watertable quickly, it could still be sufficient to produce concentrations in groundwater above 0.1 µg l^{-1}. Research activities have therefore concentrated initially on detecting pesticide residues below the soil zone, and subsequently on the mechanisms of transport and rates of degradation. If transport of pesticide residues were largely through the matrix of the UK's principal aquifers, then this would be slow and there could be time for significant attenuation by degradation and adsorption. If there was a major component of more rapid downward movement by preferential flow routes in our fractured aquifers, then pesticide residues might reach the watertable much more rapidly and without significant attenuation. Understanding the complex hydrogeology of the UK's dual porosity aquifers is thus essential to pesticide research as it has been to other groundwater resource and quality issues.

Thus, for the study of pesticides, a general strategy similar to that for nitrate was adopted. Based on considerations of usage data, observed occurrence in groundwater and physicochemical properties, compounds likely to leach to groundwater were selected for study. The approach has involved recovering undisturbed core samples, and extracting pesticide residues from both the core material and the contained water. The centrifuge extraction technique developed for nitrate has been modified for pesticides, and analytical methods for both solid material and water have been adapted or developed. In this way, the occurrence and distribution of pesticides in the unsaturated zone of the Chalk and Sherwood Sandstone aquifers have been examined. Permanent observation boreholes have been constructed to allow regular sampling immediately beneath agricultural land to which the studied pesticides have been applied in the process of normal cultivation practices (Chilton *et al.*, 1998).

The results at several sites on the Chalk aquifer indicated hardly any residues of Mecoprop or isoproturon above $0.5 \, \mu g \, kg^{-1}$ for the solid Chalk and $0.5 \, \mu g \, l^{-1}$ for the extracted porewater. Beneath fields to which atrazine had been applied to control weeds in maize cultivation, concentrations of up to $6 \, \mu g \, l^{-1}$ were detected in porewaters (Chilton *et al.*, 1993; Clark *et al.*, 1995). Regular sampling of groundwater from the uppermost part of the unsaturated zone has also detected very low concentrations of atrazine and isoproturon even when sampling has been targeted to try to observe the effects of more rapid transport.

In terms of the impact of cereal herbicides on groundwater quality in the UK, these results are generally encouraging. It appears that from normal agricultural use, very little of the applied pesticide is likely to reach the watertable. Degradation in the soil is effective in reducing concentrations, movement of very small concentrations through the matrix is slow, and the soil moisture and climatic conditions which promote preferential flow occur relatively infrequently. In the US, higher concentrations of a few micrograms per litre, but still generally not exceeding the higher health advisory levels applied there, have been observed in groundwater. These, are, however, generally in farming areas where a monoculture system means that the same compounds are used repeatedly year after year, allowing for the possibility of residues of the more persistent compounds to build up. In the UK, the combination of regular rotations in which cereals are alternated with other crops, the low pesticide application rates and the often thick unsaturated zone between cultivated soils and the water table contribute to the relatively low level of impact of pesticides from normal cultivation practices.

Nevertheless, there remain some important gaps in the overall picture. Knowledge of the fate and behaviour of pesticides for the registration process is based on studies in standard, fertile soils, and relatively little is known about the behaviour of pesticides in the deeper subsurface. From 'first principles', it can be anticipated that pesticide mobility will be greater in aquifers than in soils, because there is less organic matter and clay content for adsorption, and much lower indigenous microbial populations and nutrients for degradation (Chilton *et al.*, 1998). There is as yet little published information on subsurface pesticide behaviour. Studies underway suggest that there is indeed little scope for adsorption by aquifer materials, and recently published data (Johnson *et al.*, 1998) and continuing studies suggest that pesticide half lives could be several times longer in aquifers than in soils. The small amount of pesticide that escapes below the soil into the aquifer may persist long enough to reach the watertable, and may also persist in the saturated zone. A further uncertainty in pesticide fate is the issue of metabolites, some of which may be persistent but which are as yet often difficult to isolate analytically, and are not routinely monitored for in the UK.

7.3.6 Non-agricultural usage

As stated earlier, early monitoring programmes indicated some of the highest and most frequent detections were of pesticides which were not used agriculturally, but were employed to control weeds at locations such as railway lines, roads, car parks and other paved areas. In some of these cases, drainage of runoff by soakaways to underlying aquifers created the rapid transport routes which would allow higher concentrations to reach the watertable more rapidly. Some of these occurrences have resulted in water utilities having to invest significantly in treatment to remove pesticides at abstraction sources, and some estimates of the overall cost of this run to hundreds of millions of pounds.

The regular detection of atrazine at troublesome concentrations in areas where it was not used agriculturally has led to review of its use by the government's Advisory Committee on Pesticides, and approval for the use of atrazine in non-agricultural situations was withdrawn in 1993. There has been a rapid decline in atrazine concentrations in surface waters in response, but the much slower decline in groundwater concentrations testifies to the much greater persistence of this compound in aquifers than in soils. Many organisations replaced atrazine in their weed control programmes by other compounds such as diuron and glyphosate, and the former is beginning to appear in groundwater monitoring programmes. Other responses have included campaigns by water companies to encourage non-agricultural operators to use pesticides sparingly and carefully, and to establish localised, restricted use in sensitive areas close to public supply sources.

The few instances where much higher pesticide concentrations have been detected in groundwater can usually be attributed to point-source contamination resulting from poor handling or disposal practices or inadequate storage. In the Lincolnshire Limestone aquifer close to Peterborough, residues and waste from pesticide manufacture had been disposed of into a landfill, from which the leachate plume subsequently affected a nearby public supply source. Concentrations of mecoprop at the source exceeded 5 µg l^{-1}, requiring treatment before being use for supply, and concentrations of several hundred micrograms per litre were detected in observation boreholes close to the landfill (Sweeney *et al.*, 1998).

7.4   Hydrological change associated with mine abandonment

7.4.1   *Mining in the UK*

The UK has for many centuries sustained a major mining industry. At various periods in history, the UK has been the world's leading producer of several major minerals, including:

- tin (from Cornwall and Devon, in prehistoric times, the Roman period, and again in the nineteenth century – the last mine closed in March 1998);
- copper (mainly from Cornwall, Cumbria and Wales, in prehistoric and Roman times, then again in the eighteenth and early nineteenth centuries);
- lead (mainly from Durham, Cumbria, Wales, North Yorkshire and Derbyshire, especially in the nineteenth century);
- coal (widespread, but with the historically most productive fields in central Scotland, Durham, South Yorkshire and South Wales – the UK has been a dominant producer since the thirteenth century, with output peaking in the early twentieth century);
- oil shale (central Scotland, late nineteenth century);
- barium (mainly from Durham and Northumberland until the 1950s, and to the present day in Durham, Cumbria and Tayside).

Other minerals worked in large quantities in the UK (albeit not at world-leading levels of production) include:

- ironstone (East Pennines, Cleveland, Lincolnshire, and Northampton, with one final working mine in Cumbria);
- zinc (Cumbria, Northumberland, up to 1915);
- fluorite (Durham, Derbyshire, both still in production);
- evaporite minerals (e.g. Cheshire Salt, Cleveland Anhydrite, Cumbrian and Kentish gypsum; a few deep mines remain (one in Cheshire, one in Cumbria, one in Kent));

164

- limestone (mostly worked open-pit, with a few deep mines in the Midlands, now abandoned).

7.4.2 Hydrological impacts of working mines

The working of most of these minerals has resulted in major modifications of the hydrology of their host catchments. For instance, eighteenth and nineteenth century deep-mining of nearly all of these minerals was facilitated by the construction of long (up to several tens of kilometres) drainage adits, permanently lowering the watertable, by as much as 200 m in the North Pennines (Younger, 1998a). Such adit-drained mining districts function hydrologically as 'man-made karst' drainage systems, and indeed continue to function (or begin to flow again, where they had been later under-drained by pumping) after mine abandonment.

Temporary lowering of the watertable during mining is usually achieved by pumping, either using shafts/boreholes outside the working area, or (where the strata are less permeable) by simply collecting water in the workings in sumps (shallow basins at local low points), and then pumping it to the surface. Such mine dewatering practices can cause a regional lowering of the watertable. For instance, in the Durham Coalfield, the watertable has been lowered by more than one hundred metres beneath an area in excess of 2000 km^2 (Younger, 1993) by combined pumping of 105 Ml d^{-1} from nine pumping stations (submersible pumps in old mine shafts). The locations of these nine pumping stations are illustrated on Figure 7.7. Similarly, in the Dysart-Leven Coalfield (Fife, Scotland), an area of 50 km^2 was dewatered by two coastal shafts pumped at a total rate of 31 Ml d^{-1} (Younger et al., 1995). It is undoubtedly the case that such significant dewatering reduces natural baseflow to specific river reaches (albeit the pumped discharges may increase the flow in other reaches), and may in fact induce infiltration to the dewatered strata through permeable river beds. This is known to have been the case, for instance, at the Wheal Jane tin mine (Cornwall), where water entered the workings from the adjacent Carnon River, and at Whittle Colliery, Northumberland, where the entire summer-time flow of the Swarland Burn would disappear through the stream-bed into the limestone bed immediately overlying the worked seam.

The quality of minewater pumped from active workings is often reasonable (Younger, 1993). Indeed, minewaters have found a number of historical uses in the UK (Banks et al., 1996), ranging from mineral washing to potable supply. Some of the more mineralised minewaters have been used as sources of salt, barium and caustic soda. Even where the quality of pumped minewaters is poor, some treatment has usually been given to the waters prior to discharge, to prevent river pollution. (One notable exception to this is the discharge of untreated brines to rivers in the Midlands, where the costs of the only feasible treatment (reverse osmosis) were deemed prohibitive (Lemon, 1991)).

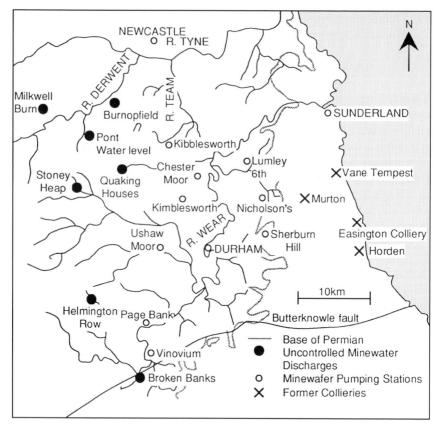

Figure 7.7 Minewater pumping stations and surface discharges in the Durham coalfield.

7.4.3 *Hydrological consequences of ceasing dewatering*

The principal hydrological consequences of ceasing dewatering are associated with the process of 'minewater rebound' (Henton, 1981; Robins, 1990; Younger, 1993; Robins and Younger, 1996), in which formerly dewatered voids gradually fill with water until a surface overflow point (often an old, shallow adit) is encountered. Figure 7.7 illustrates the locations of several surface discharges which resulted from rebound in workings to the west of the still-dewatered zone of the Durham Coalfield. Rebound not only results in a repositioning of the water table and (ultimately) surface discharges; the very process of flooding the entire volume of voids can result in a marked deterioration in minewater quality compared to that experienced during active mining. Younger (1998b) has termed this deterioration in quality a *geochemical trauma*, and it can result in the quality of surface discharges being very poor, leading to severe river pollution (Henton, 1981; Robins, 1990; Younger, 1994).

The reason why post-abandonment water quality is often so much worse than that encountered during mining can be explained as follows. Water entering drained workings does so by discrete flow-paths (often essentially as open channel flow within old workings). These flow-paths are well-washed, and any soluble minerals have usually been flushed from them. By contrast, when workings are left to flood, all of the void space is finally placed in contact with water, and all available soluble minerals dissolve rapidly. The types of mines in which this deterioration in quality is most likely to occur are obviously those in which highly soluble minerals are present. In strata which contain sulphide minerals (which readily weather to highly soluble hydroxy-sulphate minerals when exposed to air through mine ventilation), large masses of minerals may dissolve during rebound. Where the mineral pyrite (FeS_2) is involved, the dissolved products include not only iron and sulphate, but also protons (H^+), resulting in low pH. (This is due to the raising of the charge on the sulphur atoms from -1 in FeS_2 to $+6$ in sulphate). Mono-sulphide minerals (ZnS, CuS, NiS, CdS, etc.) release metals and sulphate on weathering in the presence of air, but do not generate protons since the change in the valence state of the sulphur is from -2 to $+6$, leaving no unbalanced protons. If pH is low, certain carbonate minerals (such as siderite ($FeCO_3$) and rhodocrosite ($MnCO_3$)) can also dissolve rapidly, releasing more metals into solution. Indeed, some workings containing siderite but virtually no sulphide generate ferruginous drainage by direct dissolution of siderite by infiltrating waters. (Further details on minewater chemistry evolution can be found in the recent paper by Banks *et al.*, 1997.)

As these sulphide and carbonate minerals are widely disseminated in mined strata in Britain, workings of all the minerals listed in section 7.1 (with the exception of limestone) can result in discharges of acidic and/or metalliferous waters to rivers, resulting in serious environmental degradation (e.g. NRA, 1994; Jarvis and Younger, 1997). However, each case must be evaluated on its own merits, for not all mines working the same mineral have the same pollution potential. For instance, the Wheal Jane tin mine (closed March 1991, dewatering ceased on 6 June 1991) produced the most extremely polluted discharge in British history, whereas the nearby Geevor tin mine (which closed in June 1991, and began to flow at surface in January 1995) produced water of good quality. All indications are that the recently-closed South Crofty tin mine (which lies geographically between Geevor and Wheal Jane) will be much more like Geevor than Wheal Jane in terms of future discharge water quality.

The process of minewater rebound can have further consequences beyond river pollution (Younger, 1998c). These include:

- localised flooding of agricultural and industrial land (e.g. at St Helen Auckland, County Durham; Younger, 1998c);
- temporary loss of dilution of sewage effluents in the period between cessation of pumping and commencement of surface discharges (Banks *et al.*, 1996);
- over-loading and clogging of sewers which is a major problem in the abandoned Cleveland Ironstone field;
- pollution of over-lying aquifers by upward migration of minewater, which is known to have occurred in the Magnesian Limestone aquifer near Bishop Auckland, Co. Durham;
- temporarily accelerated mine gas emissions, especially in relation to dense gases such as carbon dioxide and radon which can literally 'ride up' on the rising watertable;
- the increased risk of subsidence as rising waters weaken support pillars, and possibly as faults are reactivated (L. Donelly, British Geological Survey, personal communication).

7.4.4 Issues of time scale

The time lag between cessation of dewatering and onset of surface discharges varies dramatically between different mine systems. For instance, Geevor and Wheal Jane mines were abandoned within three months of each other. Geevor did not give rise to a surface discharge for 42 months, whereas Wheal Jane had flooded to ground level within nine months. Since both mines are in west Cornwall, and thus subject to similar rainfall patterns, the differences must relate to either:

- the amount of rainfall which actually becomes recharge; and/or
- the volume of mined voids and drained strata which must be flooded.

In this particular case, it is interesting to note that the most recent Geevor workings were under the seabed (where great precautions are taken to minimise inflows from above) whilst Wheal Jane workings are not only inland, but exploited a lode which passes right through a major river valley.

Observations elsewhere in the UK have shown that in large regional coalfields, a decade or more may be required before rebound is complete. Monitoring in the main Northumberland Coalfield indicates that 17 years will have elapsed between the cessation of dewatering and the first surface discharges (expected around 2003). Model predictions for other coalfields indicate even greater time lags, of 18 years (Dysart-Leven Coalfield; Younger *et al.*, 1995) or even 40 years (Durham Coalfield; Sherwood and Younger, 1994). Active rebound in regional coalfields is currently underway in Lancashire, Derbyshire, South Yorkshire and Kent, the consequences of which may not be experienced until well into the twenty-first century.

Plate 7.1 Iron-rich minewater discharge to the River Esk below Dalkeith Palace (Nick Robins).

Once discharges at surface commence, a different question of time scale arises: the time required for the quality of the discharging minewater to improve to long-term levels. The UK is a particularly useful study area for investigating this question, for there are numerous abandoned mine discharges which we know to be coming from mines in which pumping was discontinued at various periods before present (some more than a century ago). Younger (1997a) has described the long-term evolution of minewater quality after the commencement of surface discharge, noting that an exponential decline in contaminant concentrations is the norm. The general pattern is for the long-term concentration of iron to be around 10% of the initial concentration, and for this long-term concentration to be reached after a period of time equal to about five times the period between cessation of pumping and first surface discharge. Thus a minewater, which took one year to reach the surface, and contains 250 mg l^{-1} of iron initially, will have 25 mg l^{-1} of iron five years after flow commenced. The 25 mg l^{-1} concentrations will then persist until source minerals in the zones of active dissolution (typically where the water table fluctuates in shallow workings) are finally exhausted. Calculations to date suggest that exhaustion of these source minerals may require several centuries (Banwart, personal communication, 1998).

7.4.5   *Prevention and remediation strategies*

Strategies for preventing and remediating these various impacts are undergoing rapid development in the UK. For surface water pollution problems, existing technology from the mining industry (involving alkali dosing and sedimentation) is being applied at several sites (most notably Wheal Jane). At around a dozen abandoned mine sites, more novel, passive treatment techniques are being implemented. These primarily involve the construction of various kinds of wetlands, which are essentially geochemical reactors, configured to maximise removal of acidity, iron and other metals. A full account of UK experiences to date is given in Younger (1997b).

Prevention of future problems is being ensured in several English coalfields (West Yorkshire, Staffordshire and Durham) by the Coal Authority, who currently operate 11 pumping stations (nine of which are in Durham) designed to prevent rebound of minewaters to the land surface. Similarly, in Scotland, a scheme to prevent predicted pollution of the River Almond by minewater currently nearing surface in the Polkemmet area of West Lothian (Chen *et al.*, 1997) has prompted the Coal Authority to implement a pump-and-treat scheme, which was commissioned in April 1998. A similar scheme is also planned by the Coal Authority for the Dysart-Leven Coalfield (Fife), to prevent pollution of the River Leven which was predicted by Younger *et al.* (1995). Regulatory understanding and analysis of such preventative pumping systems has required the development of new modelling approaches. This is because existing groundwater models, which assume laminar flow in intergranular porous media are not very appropriate for conditions in deep mines, where turbulent flow in open roadways can be a major component of the groundwater flow system. Two new approaches to modelling mined systems have been developed. The simplest of these is a semi-distributed model (GRAM) which represents discrete bodies of workings as linear reservoirs connected by pipes (Sherwood and Younger, 1997). A more comprehensive representation of heterogeneities in mined systems is possible using a physically-based modelling approach, in which a three-dimensional pipe network model (representing major roadways, etc.) is routed through a three-dimensional variably saturated porous medium model (representing areas of collapsed workings and *in situ* strata (Adams and Younger, 1997).

7.5   Urbanisation and groundwater

7.5.1   *Inputs to groundwater: recharge*

Groundwater is usually replenished by direct recharge, which is the deep percolation of excess rainfall through the soil to the aquifer below. In cities, however, much of the surface is impermeable, with as much as 80% of city centres being covered by buildings, roads and carparks. This proportion reduces to 10% or so in green, low-density suburbs. Impermeabilisation of

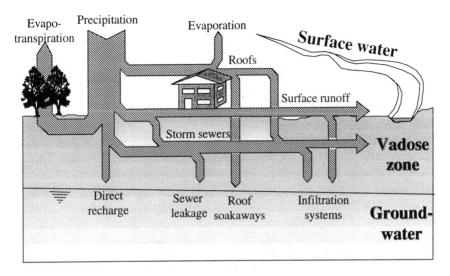

Figure 7.8 Urban pathways from precipitation to groundwater recharge. (After Lerner, 1997.)

the surface is bound to decrease direct recharge by diverting water away from the soil, as illustrated in Figure 7.8. There remains some debate about whether a significant part of this diverted water finds its way to groundwater by other routes such as soakaways, leaking drains, or deliberate infiltration of excess stormwater (Lerner, 1997). The truth is that these alternative routes have varying significance between cities, depending on local geology and the city infrastructure. For example, the use of boreholes to infiltrate road runoff in cities in rare in the UK, but does occur in Norwich.

Cities import water for public supply, i.e. for supply through a network of water mains to domestic, commercial and industrial users. The imports can be of surface water from reservoirs (e.g. Birmingham), from rivers (Coventry), or of groundwater from surrounding rural areas (Nottingham). They are sometimes supplemented by urban groundwater for public supply (Coventry) or industrial use (Birmingham, Nottingham). Annual imports are comparable in size to annual rainfall, when both are expressed in the same units as a depth of water spread evenly over the area of the city. For example, Birmingham imports about 600 mm y^{-1}, which is almost identical to the average rainfall of 700 mm y^{-1}. Water mains are well known to leak, and loss rates of 25% are commonplace in the UK. The lowest rate is reported to be for Hull, at 8% (personal communication, Gert Cashandt, Yorkshire Water). As shown schematically in Figure 7.9, much of this leakage will recharge groundwater because it has been introduced into the soil below the impermeable surfaces of the city and below the influence of most plants. Some of the leakage will be captured by sewers, where these lie beneath the water mains.

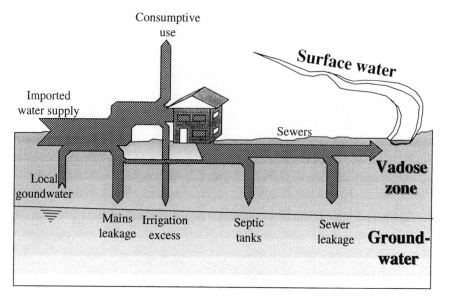

Figure 7.9 Pathways for water supply to waste and groundwater recharge. (After Lerner, 1997.)

A second network of water-carrying pipes exists in cities – the sewers. Do these leak significantly? The evidence is that they do, but in both directions. When a sewer is below the watertable, it is likely to gain water because it is unpressurised in contrast to a water main, which is usually at a large positive pressure. Sewers above the water table can lose water, which can recharge groundwater. Leakage rates are probably not as high as for water mains because of the differences in pressures, and the net effect is mainly on groundwater quality. Recent research by Yang *et al.* (submitted) suggests that the deterioration of groundwater quality under Nottingham due to leaking sewers is equivalent to a net outward leakage of 10 mm y^{-1} over the city.

The net effect of impermeabilisation, storm water disposal, mains and sewer leakage seems to be no change in total recharge in UK cities. Greswell *et al.* (1994) showed this for Birmingham, where they calculated the change in recharge from 1900–89 was only a decrease of 4%, well within the margins of error of such a calculation. Similar results were reported for Liverpool (Rushton *et al.*, 1988) and Nottingham (Yang *et al.*, submitted). The latter case is interesting in that it is the first study in which the proportions of recharge from each of the three principal sources (rainfall, mains, sewers) has been quantified for a UK city, as shown in Figure 7.10. The proportion of recharge that comes from rainfall has decreased dramatically over the 150 years of the study period.

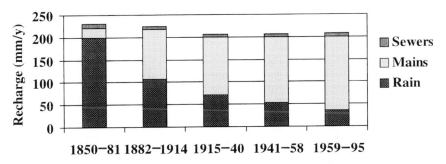

Figure 7.10 Estimated recharge from rainfall, leaking water mains and leaking sewers in Nottingham city centre. (After Yang *et al.*, submitted.)

7.5.2 *Outputs: pumping*

The natural discharge for groundwater is to rivers and streams. However, groundwater is such a convenient source of water that it has been heavily exploited by pumping from wells, particularly in urban areas. Similar patterns of abstraction have been experienced in most of the industrial cities of the UK. Significant abstraction began in the early to mid-eighteenth century with steam-powered pumping stations taking water for individual factories and public supply, with the wells for the latter often being located outside of the towns. Towns grew to enclose many of the public supply wells, and demand increased. Industrial activity continued to expand and many factories and breweries (Lloyd, 1986) had their own wells. This pattern of growth continued up to the Second World War and into the 1950s. Since then, there have been major changes in British industry. Much heavy industry has closed, and large engineering factories have downsized, relocated or split into smaller units, with a consequent decline in urban groundwater pumping. Figure 7.11 illustrated this pattern for Birmingham. Pumping grew steadily until the 1950s, and probably exceeded recharge to the aquifer by 1900. The continued growth was despite the shift away from groundwater for public supply with the advent of the Elan Valley scheme that brought surface water from Wales to Birmingham in 1901. The watertable fell by up to 50 m by 1966 (Knipe *et al.*, 1993). However, by 1990, abstraction had fallen to 20% of its peak and is now less than recharge.

While pumping from urban aquifers was high, the natural discharge to rivers was sometimes reversed, with recharge of the aquifer by the river in areas where the watertable was lowered. In coastal areas, the drawdown of the watertable sometimes induced saline water to intrude in to the aquifer, e.g. in Liverpool. These problems have occurred in many other countries. Subsidence of the ground has been another effect of falling watertables where there are compressible strata (clays). Fortunately, this has not been a serious problem in the UK.

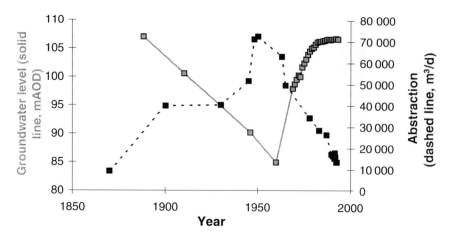

Figure 7.11 Birmingham aquifer: comparison of total abstraction rate and groundwater levels at Constitution Hill observation borehole. (After Greswell
et al., 1994.)

7.5.3   *Effects: rising groundwater*

The overall effect of the changes to recharge and pumping in UK cities has
been to cause watertables to rise in recent years. Brassington (1990) reviewed
the known cases and identified problems in 18 locations in the UK and a
number of cases overseas. Further overseas examples can be found in Wilkinson
(1994) and Chilton *et al.* (1997). Selected examples are summarised in
Table 7.1.

One of the cities in which rising groundwater problems were first recognised and acted upon is Birmingham (Knipe *et al.*, 1993; Greswell *et al.*, 1994).
The fall in abstractions since the 1950s has been mirrored by a rise in the
watertable (Figure 7.11). By the late 1970s, groundwater levels were returning to near-surface in the natural discharge area of the Tame valley. Machine
pits and basements in factories were being flooded. A group of 20 m deep
wells was installed to pump groundwater with the sole purpose of keeping
the watertable below factory operations. Detailed studies of the Birmingham
aquifer were undertaken in the early 1990s and included surveys that showed
that some minor damage to the foundations of houses could be attributed
to the regional rise of the watertable. A computer model of the groundwater system was constructed. It predicted that most of the rise was complete
in critical areas where damage or flooding would result, but rises would
continue under the city centre and further south where they were not
critical.

The consequences of rising groundwater under London are potentially
more serious (Simpson *et al.*, 1989). The Chalk aquifer under central London

Table 7.1 Selected examples of rising groundwater in the UK
(based on Brassington, 1990)

Location	Rise in groundwater levels	Reduction in abstraction (Ml/d = 10^6 litres per day)	Problems caused
London	20 m at Trafalgar Square, 1965–86	1940–240 Ml/d 1982–118 Ml/d	Potential foundation problems, flooding of underground
Wolverhampton	18 m over period 1973–87	Closure of 20 Ml/d factory well	None reported
Fawley (Hants)	15–45 m over period 1950–85	1950–4.5 Ml/d 1981–0 Ml/d	None reported
Doncaster	Not measured	'considerable'	Sewer flows increased as groundwater intercepted
Liverpool	2.5 m over period 1975–85	1963–45 Ml/d 1984–10 Ml/d	Flooding in rail tunnels and some basements, pumping needed in tunnels

is relatively deep below the ground, and is overlain by London Clay (and other strata). Large buildings in London have a variety of foundation designs to secure them in the softer clay, including concrete rafts, end bearing piles, piles that rely on friction along their shafts, and large basements that 'float' the building on the clay. Lowering groundwater levels beneath the clay has encouraged slow drainage and a lowering of porewater pressures and generally increased the strength of the clay. Rising groundwater and porewater pressures can lower strengths again, and cause swelling. Although no buildings are likely to fail, damage may well occur, especially when two parts of a building are on different types of foundation. A scheme is being considered which will control groundwaters by pumping for a number of wells, possibly using the water for factories and offices (personal communication, Phil Aldous, Thames Water).

Cities have huge impacts on their underlying groundwater. Not only does it become polluted, but the amount and source of recharge changes dramatically. Traditionally urban groundwater has been over-exploited for supply, but this has changed in recent years in the UK, with the result that water tables are rising. This in turn can have undesirable results, ranging from increased flow in sewers and flooding of basements, to severe structural a damage to buildings as their foundations are affected.

175

References

Adams, R. and Younger, P.L. (1997) Simulation of groundwater rebound in abandoned mines using a physically based modelling approach, in Veselic, M. and Norton, P.J. (eds), *Proceedings of the 6th International Mine Water Association Congress, Minewater and the Environment*, Bled, Slovenia, 8–12 September 1997. Vol. 2, pp. 353–62.

Addiscott, T.M., Whitmore, A.P. and Powlson, D.S. (1991) *Farming, Fertilisers and the Nitrate Problem*. CAB International, Wallingford. p. 170.

Banks, D., Younger, P.L., Arnesen, R.-T., Iversen, E.R. and Banks, S.D. (1997) Minewater chemistry: the good, the bad and the ugly. *Environ. Geol.*, **32**(3), 157–74.

Banks, D., Younger, P.L. and Dumpleton, S. (1996) The historical use of mine-drainage and pyrite-oxidation waters in central and eastern England, United Kingdom. *Hydrogeol. J.*, **4**(4), 55–68.

Brassington, F.C. (1990) Rising groundwater levels in the United Kingdom. *Proc. Inst. Civ. Eng.* I, **88**, 1037–57.

Burston, M.W. Nazari, M.M., Bishop, P.K. and Lerner, D.N. (1993) Pollution of groundwater in the Coventry region (UK) by chlorinated hydrocarbon solvents. *J. Hydrol.*, **149**, 137–61.

Burt, T.P., Heathwaite, A.L. and Trudgill, S.T. (1993) *Nitrate: Processes, Patterns and Management*. Wiley, Chichester. p. 444.

Calder, R. (1992) Hydrologic effects of land use change, in Maidment, D.R. (ed.) *Handbook of Hydrology*. McGraw-Hill, New York.

Chen, M., Soulsby, C. and Younger, P.L. (1997) Predicting water quality impacts from future minewater outflows in an urbanized Scottish catchment, in Chilton, P.J. *et al.* (eds), *Groundwater in the Urban Environment, Vol. 1: Problems, Processes and Management*, Proceedings of the XXVII Congress of the International Association of Hydrogeologists, Nottingham, UK, 21–27 September 1997. Balkema, Rotterdam. pp. 383–8.

Chilton, J. *et al.* (eds) (1997) *Groundwater in the Urban Environment, Vol. 1; Problems, Processes and Management*, Balkema, Rotterdam.

Chilton, P.J., Lawrence, A.R. and Stuart, M.E. (1998) Pesticides in groundwater: some preliminary results from recent research in temperate and tropical environments, in Mather, J., Banks, D., Dumpleton, S. and Fermor, M. (eds), *Groundwater Contaminants and their Migration*. Geological Society of London, Special Publication 128. pp. 333–45.

Chilton, P.J., Stuart, M.E., Gardner, S.J., Hughes, C.D., Jones, H.K., West, J.M., Nicholson, R.A., Booker, J.A., Bridge, L.R. and Goody, D.C. (1993) Diffuse pollution from land-use practices. National Rivers Authority. Project record 113/10/ST.

Clark, L., Turrell, J., Fielding, M., Oakes, D.B., Wilson, I. and Taylor, L. (1995) Pesticides in major aquifers. *National Rivers Authority R&D Report 17*. National Rivers Authority, Bristol.

DoE (1995) *Digest of Environmental Statistics No. 17, 1995*. HMSO, London.

DoE (1996) *Indicators of Sustainable Development for the UK*. HMSO, London.

Downing, R.A. (1993) Groundwater resources their development and management in the UK: an historical perspective. *Quart J. Eng. Geol.*, **26**, 335–58.

Downing, R.A. (1998) *Groundwater – Our Hidden Asset*. British Geological Survey, Nottingham.

Ford, M. and Tellam, J.H. (1994) Source, type and extent of inorganic contamination within the Birmingham urban aquifer system, UK. *J. Hydrol.*, **156**, 101–35.

Foster, S.S.D., Bridge, L.R., Geare, A., Lawrence, A.R. and Parker, J.M. (1986) The groundwater nitrate problem. British Hydrology Research Report 86/2.

Foster, S.S.D., Chilton, P.J. and Stuart, M.E. (1991) Mechanisms of groundwater pollution by pesticides. *J. Inst. Water Environ. Manag.*, **5**, 1986–193.

Greswell, R.B., Lloyd, J.W., Lerner, D.N. and Knipe, C.V. (1994) Rising groundwater in the Birmingham area, in Wilkinson, W.B. (ed.), *Groundwater Problems in Urban Areas*, Thomas Telford, London. pp. 330–41, discussion pp. 355–68.

Henton, M.P. (1981) The problem of water table rebound after mining activity and its effects on ground and surface water quality, in van Duijvenbooden, W., Glasbergen, P. and van Lelyveld, H. (eds), *Quality of Groundwater*, Proceedings of an International Symposium, Noordwijkerhout, The Netherlands, held 23–27 March 1981. Elsevier, Amsterdam. pp. 111–16.

House of Lords (1989) Nitrate in Water. Select committee on the European Community. HMSO, London. p. 274.

Jarvis, A.P. and Younger, P.L. (1997) Dominating chemical factors in mine water induced impoverishment of the invertebrate fauna of two streams in the Durham Coalfield, UK. *Chem. Ecol.*, **13**, 249–70.

Johnson, A.C., Hughes, C.D., Williams, R.J. and Chilton, P.J. (1998) Potential for aerobic isoproturon biodegradation and sorption in the unsaturated and saturated zones of a chalk aquifer. *Journal of Contaminant Hydrology*, **30**. Elsevier, Netherlands. pp. 281–297.

Knipe, C.V., Lloyd, J.W., Lerner, D.N. and Greswell, R.B. (1993) Rising groundwater in Birmingham and the engineering implications. *CIRIA Special Publication 92*. London.

Lemon, R. (1991) Pumping and disposal of deep strata mine water. *Mining Technol.*, March 1991, 69–76.

Lerner, D.N. (1996) Urban groundwater – an asset for the sustainable city? *European Water Pollut. Cont.*, **6**(5), 43–51.

Lerner, D.N. (1997) Too much or too little – recharge in urban areas, in Chilton, J. *et al.* (eds), *Groundwater in the Urban Environment, Vol. 1; Problems, Processes and Management*, Balkema, Rotterdam. pp. 41–7.

Lloyd, J.W. (1986) Hydrogeology and beer. *Proc. Geol. Assoc.*, **97**, 213–19.

NERC (1993) *Hydrological Data United Kingdom: Hydrometric Register and Statistics 1986–90*. Natural Environment Research Council. Institute of Hydrology, Wallingford.

NRA (1993) *Low Flows and Water Resources: Facts on the Top 40 Low Flow Rivers in England and Wales*, National Rivers Authority, Bristol.

NRA (1994) Abandoned mines and the water environment. *Report of the National Rivers Authority. Water Quality Series No 14*. London, HMSO.

Rivett, M.O., Lerner, D.N., Lloyd, J.W. and Clark, L. (1990) Organic contamination of the Birmingham aquifer. *J. Hydrol.*, **113**, 307–23.

Robins, N.S. (1990) *Hydrogeology of Scotland*. HMSO for British Geological Survey, London.

Robins, N.S. and Younger, P.L. (1996) Coal abandonment – mine water in surface and near-surface environment: Some historical evidence from the United Kingdom, in *Proceedings of the Conference on Minerals, Metals and the Environment II*, Prague, Czechoslovakia, 3–6 September 1996. Institution of Mining and Metallurgy, London. pp. 253–62.

Rushton, K.R. *et al.* (1988) Groundwater model of conditions in the Liverpool sandstone aquifer. *J. Inst. Water Environ. Manag.*, 2, 67–84.

Sherwood, J.M. and Younger, P.L. (1994) Modelling groundwater rebound after coalfield closure: An example from county Durham, UK, in *Proceedings of the 5th International Minewater Congress*, Nottingham, UK. Vol. 2, pp. 769–77.

Sherwood, J.M. and Younger, P.L. (1997) Modelling groundwater rebound after coalfield closure, in Chilton, P.J. *et al.* (eds), *Groundwater in the Urban Environment, Vol. 1: Problems, Processes and Management*, Proceedings of the XXVII Congress of the International Association of Hydrogeologists, Nottingham, UK, 21–27 September 1997. Balkema, Rotterdam. pp. 165–170.

Simpson, B., Blower, T., Craig, R.N. and Wilkinson, W.B. (1989) The engineering implications of rising groundwater levels in the deep aquifer beneath London. *CIRIA Special Publication 69*. London.

Sweeney, J., Hart, P.A. and McConvey, P.J. (1998) Investigation and management of pesticide pollution in the Lincolnshire Limestone aquifer in eastern England, in Mather, J., Banks, D., Dumpleton, S. and Fermor, M. (eds), *Groundwater Contaminants and their Migration.* Geological Society of London, Special Publication 128, pp. 347–60.

Wilkinson, W.B. (ed.) (1994) *Groundwater Problems in Urban Areas.* Thomas Telford, London.

Yang, Y., Lerner, D.N., Barrett, M.H. and Tellam, J.H. (submitted). Quantification of groundwater recharge in the city of Nottingham, UK.

Younger, P.L. (1993) Possible environmental impact of the closure of two collieries in County Durham. *J. Inst. Water Environ. Manag.*, 7, 521–31.

Younger, P.L. (1994) Minewater pollution: the revenge of old king coal. *Geoscientist*, 4(5), 6–8.

Younger, P.L. (1997a) The longevity of minewater pollution: a basis for decision-making. *Sci. Total Environ.*, 194/195, 457–66.

Younger, P.L. (ed.) (1997b) *Minewater Treatment Using Wetlands*. Proceedings of a National Conference held 5 September 1997, at the University of Newcastle, UK. Chartered Institution of Water and Environmental Management, London.

Younger, P.L. (1998a) Adit hydrology in the long-term: observations from the Pb–Zn mines of northern England, in *Proceedings of the International Mine Water Association Symposium, Mine Water and the Environment*, Johannesburg, South Africa, 6–10 September 1998. International Mine Water Association, Johannesburg.

Younger, P.L. (1998b) Coalfield abandonment: geochemical processes and hydro-chemical products, in Nicholson, K., *Energy and the Environment. Geochemistry of Fossil, Nuclear and Renewable Resources.* Society for Environmental Geochemistry and Health. McGregor Science, Aberdeenshire. In press.

Younger, P.L. (1998c) Hydrological consequences of the abandonment of regional mine dewatering schemes in the UK, in *Proceedings of the International Symposium on Hydrology in a Changing Environment.* Organised by the British Hydrological

Society, 6–10 July 1998, Exeter, UK. British Hydrological Society Occasional Paper 9, pp. 80–82. British Hydrological Society, Wallingford.

Younger, P.L., Barbour, M.H. and Sherwood, J.M. (1995) Predicting the Consequences of Ceasing Pumping from the Frances and Michael Collieries, Fife, in Black, A.R. and Johnson, R.C. (eds), *Proceedings of the Fifth National Hydrology Symposium, British Hydrological Society*. Edinburgh, 4–7 September 1995. pp. 2.25–2.33. British Hydrological Society, Wallingford.

8

LAKES AND PONDS

Tony Bailey-Watts, Alex Lyle, Rick Battarbee,
Ron Harriman and Jeremy Biggs

This chapter concerns the many thousands of standing waters, ranging from tiny pools to massive systems such as Lough Neagh, Northern Ireland and the large Scottish Lochs, e.g. Ness, Morar and Awe. It focuses on (1) water resources, dynamics and ecological effects, (2) ponds as a 'special' category of lentic waters, and (3) the main 'threats' to the well-being of these systems, i.e. climate change, acidification and accelerated nutrient enrichment (eutrophication). Each of the contributions reflects the holistic and inter-related nature of the subject of standing waters.

8.1   Introduction

8.1.1   Resources

Quantification of the total UK lake resource will be imprecise. This is mainly due to ambiguity of definition at the lower size limits, e.g. the division between lakes and ponds and the lower size limits for inclusion on maps. Furthermore, some of the smallest waters are ephemeral. From a survey of fresh waters in Great Britain (GB) (Smith and Lyle, 1979), the number of standing waters on maps of 1:250 000 scale was 5505; the minimum size being approximately 4 ha. Extrapolation of numerical comparisons for sample areas between 1:250 000 and 1:63 360 scale maps suggests that some 81 000 waters are shown on the latter. In a separate study for Northern Ireland, Smith *et al.*, (1991) counted 1668 waters on 1:50 000 scale maps, the lowest size category being <0.25 ha. The percentages of land area covered by inland waters in the four countries of the UK are: 0.52% in England; 0.60% in Wales; 2.04% in Scotland (all from Central Office of Information, 1977); and 4.43% in Northern Ireland (Smith *et al.*, 1991). The percentages for GB and UK are 1.04% and 1.24% respectively. The distribution of lakes throughout GB (by hydrometric area) is demonstrated in Figures 8.1(a) and 8.1(b) which indicate the number per unit area, and the percentage area of standing water (from Smith and Lyle, 1979).

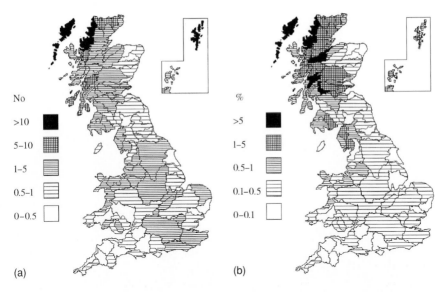

Figure 8.1 (a) The distribution of lakes in Great Britain by hydrometric area, numbers per 100 km² (from Smith and Lyle, 1979). (b) The percentage area of standing water in each hydrometric area in Great Britain (from Smith and Lyle, 1979).

Smith and Lyle (1979) analysed the numbers of GB lakes in nine logarithmic size classes, i.e. 0.25–0.50 km², 0.50–1.0 km², etc., which demonstrated a declining trend in numbers with increasing size categories. Scottish lochs show a strong relationship in this respect where the correlation coefficient (r) is 0.994 (Lyle and Smith, 1994). However, in Northern Ireland there is a bias to small waters where about 75% are less than 2 ha. The accumulated areas of standing water in GB in each of the nine size categories above were broadly similar (mean = 136.8 km², SD = 34.1 km²) but in Northern Ireland the five largest waters represent 90% of the total lake surface area. The largest lake there, and also in the UK, is Lough Neagh at 380 km². The distribution of lake water by volume is strongly biased towards the few large and deep lochs in Scotland (Smith and Lyle, 1979; Lyle and Smith, 1994 (Plate 8.1)). For example, the volume of Loch Ness alone accounts for almost 20% of the estimated GB total, and holds one and half times the water of Lough Neagh yet covers only one-seventh of its area. Smith and Lyle (1979) estimated the total water volume in the 5505 counted lakes ≥4 ha to be 38.3 km³.

Whether the national resource in terms of numbers of standing waters is increasing or decreasing is not recorded. On one hand, many lakes have been lost naturally over time by siltation and some smaller waters have been filled or drained by man. On the other, new or expanded waters have been created artificially, some large reservoirs for power and water supply, and many

Plate 8.1 Loch Tummel, Perthshire (Mike Acreman).

smaller ones for industrial and recreational purposes. Smith and Lyle (1979) estimated that about 10% of the 5505 waters counted were reservoirs. However, the overall area of standing water has almost certainly increased in recent times under the influence of man. For example, during this century in Scotland, 28 reservoirs, all >2 km², have been created adding 115 km² (approximately 7.5%) to the total lake surface area there (Lyle and Smith, 1994).

8.1.2 *Dynamics and ecological effects*

The climatic forces which drive lake dynamics are rainfall, wind and solar radiation. Their effects are interactive and are briefly described below.

8.1.3 *Hydrology*

Clearly, rain is fundamental in providing water to form lakes by way of runoff from their catchments. Runoff is the net result of rainfall less evaporation, and the latter is influenced by winds and solar radiation. Catchment size, topography, geology, soils, vegetation and land use are all influential in determining the quantities and the rates at which runoff enters the lakes. The most obvious feature of runoff fluctuation is change in lake surface level. Lakes with stable water levels tend to lie in porous catchments where a high proportion of the inflow comes from ground storage. Short-term rain events are minimally reflected in lake level and long-term reductions in average rainfall would lead to consistently lower levels. In contrast, lakes on impervious catchments respond quickly to rain events and generally experience a greater range and frequency of change in water level. The ecological status of lakes is partially characterised by the effects of water level stability. Lakes with a narrow level range will have a greater proportion of silty substrates in the shoreline margins than those with pronounced and frequent changes, which in turn, generally have stone/gravel shore areas. These contrasts will be reflected in their plant and animal communities. For example, the former may support stands of emergent vegetation which could not survive the periodic desiccation experienced in the latter and, similarly, invertebrate fauna will differ between these substrate types, as will fish species because of their different spawning habitat requirements.

The rate of water flow through lakes clearly affects the hydrological dynamics within them, but also their chemical and biological status. The water replacement time influences biological responses to nutrient supplies, e.g. if less than the generation or doubling time of planktonic algae (Bailey-Watts et al., 1992), and pollutants, e.g. as with acidification influxes (Lyle, 1987). Furthermore, in lakes which are thermally stratified in the summer months, the water volume being replaced is only that of the upper warmer layer (the epilimnion, see below) thus reducing the effective replacement time considerably. Through-flow is therefore a useful parameter in lake classification, and ranges from a few days (perhaps only hours during floods) in some small lakes, to several years in large waters, e.g. 6.9 years for Loch Morar (Smith et al., 1981). Hence, changes in the temporal patterns of rainfall (and evaporation) will alter the physical, chemical and biological behaviour of lakes.

8.1.4 Mixing

Wind stress at the lake surface is the dominant force affecting water movement and mixing within the lake basins. Waves, circulation patterns and oscillations (seiches) are created by winds, their scale and form being dependent on wind speed, direction and duration, and the morphometric characteristics of the lake, i.e. shape, area (fetch) and bathymetry. The currents and turbulence created are also influential on the chemical and biological status of lakes. In shallow waters, strong wind-induced turbulence which reaches and disturbs the lake bed will increase turbidity and release and mix nutrients from the sediments into the water column. Turbidity will reduce light penetration which affects photosynthetic phytoplankton and productivity, whereas released nutrients could promote productivity. While some wind mixing and circulation is necessary to create a stable thermally stratified structure in deeper lakes (see below), very strong winds may disrupt this structure and force mixing between the epilimnion and the cooler (and probably chemically and biologically different) lower layer (the hypolimnion).

The stability of shoreline features, i.e. substrate and vegetation, can be temporarily destroyed by large waves breaking onto shores thus damaging the habitats of invertebrates and the spawning areas of some fish species. Consequently, increased frequency of extreme wind events, i.e. calms and storms, may significantly alter the ecological character of many lakes.

8.1.5   Thermal characteristics

Solar radiation and thermal conductivity at the air/water interface are the principal sources of heat exchange governing lake temperature. Most UK lakes can be classified as dimictic, i.e. they fall below the temperature of maximum density (4°C) in winter and rise above that in summer (Hutchinson, 1957). A few of the largest Scottish lochs are warm monomictic, i.e. they do not fall below 4°C. In such lochs, the winter dissipation of heat energy accumulated throughout the summer can noticeably influence the immediate local climate. Temperature is of fundamental importance to the species composition of the flora and fauna found in any lake. While short-term deviations from the 'normal' heating and cooling sequences may be tolerated by most plants and animals, prolonged and consistent change in the thermal regime will effect a change in biological status.

Thermal stratification is one of the more important physical features to occur in lakes which are deep enough to support it (e.g. Gorham and Boyce, 1989) and is also a feature significant to lake ecology (Golterman, 1975). Summer heating and wind mixing of the surface waters creates a relatively buoyant warm layer (the epilimnion) which is effectively cut off from the cooler, lower water (the hypolimnion) by the thermocline. The creation of such a thermal and hydraulic structure is important ecologically in that the epilimnion is a zone promoting biological productivity in contrast to the cooler, darker and possibly poorly oxygenated hypolimnion.

In winter, once the water column temperature has cooled below 4°C, surface ice may form during sub-zero air temperatures. With the probable exception of smaller high-altitude lakes, a regular and prolonged winter freeze is not a significant feature of UK lakes. Most of the larger deep lakes seldom have any ice. However, some other lakes currently experience sufficiently long periods of freezing to effect an ecological impact. Light, oxygen and nutrients are required to sustain biological productivity and all three are affected by ice cover. Solar radiation can be blocked by opaque or snow-covered ice. Turbulence by winds to aerate the water stops and existing dissolved oxygen is depleted by biological functioning and decay. Also, the resuspension of nutrients from sediment disturbance by wind action ceases. This can create an inhospitable environment for life and in extreme cases fish may die through suffocation. Even minor cooling of average winter temperatures will significantly increase the frequency and duration of lake ice throughout the UK with important ecological effects. Analysis of records for Loch Leven (e.g. Lyle, 1981) shows that a fall in mean winter air temperature from 4°C to 2°C gives a five-fold increase in the number of days when ice is present.

8.2 Pond ecology

8.2.1 *Introduction and status of ponds*

Ponds include a wide range of man-made and natural water bodies. In Britain they have been defined as small water bodies between 1 m^2 and 2 ha in area which hold water for four months of the year or more (Biggs *et al.*, 1996).

Ponds are often thought of as quintessentially artificial habitats but the geological record shows that they have always been a natural feature of landscapes; however, human activity has added a variety of new ways in which ponds can be formed. Consequently, it is informative to think of ponds as a natural habitat type, which has often been recreated by human activity. It should be noted that, even before landscapes were widely influenced by anthropogenic processes, the pond environment was probably very common, with a wide range of natural processes capable of creating ponds (Biggs *et al.*, 1994).

Until recently, comparatively little has been known about the ecology of ponds. It seems likely that the small size of ponds has led to them being regarded as generally unimportant ecologically. Increasingly, however, it is becoming clear that small water bodies are biologically rich, and data from the UK have been valuable in altering our perspective on the ecology of ponds. Three studies have been particularly important in this respect: the National Pond Survey initiated by Pond Action in 1990 with support from World Wildlife Fund, UK; the Natural Environment Research Council (NERC)-funded work on impaired ponds throughout England and Wales; and the DETR (Department of the Environment, Transport and the Regions) Lowland Pond Survey 1996 (Williams *et al.*, 1998).

Collectively, these studies provide clear evidence of the biological richness of small water bodies. For example, recent comparisons of aquatic macro-invertebrate species richness in rivers and ponds in the UK show ponds to support more species, and more uncommon species of conservation import-ance, than rivers (Table 8.1). Similarly, the DETR Lowland Pond Survey showed that just 377 lowland ponds accounted for 50% of Britain's wetland plants. In addition, more freshwater species scheduled on the UK Wildlife and Countryside Act and listed in the UK Biodiversity Action Plan shortlist are found in ponds than either rivers or lakes (Williams *et al.*, 1998).

World-wide, small water bodies occur in all landscape types from mountain tops to deserts. In the highly modified British landscape ponds are most

Table 8.1 Comparison of macroinvertebrate biodiversity in ponds and rivers in the United Kingdom: species richness and occurrence of uncommon species.

	Species richness		Species richness spp.		Nationally scarce Red Data Book spp.	
	Rivers	Ponds	Rivers	Ponds	Rivers	Ponds
Flatworms	8	9	1	0	0	0
Snails and orb mussels	34	33	1	2	4	2
Leeches	10	14	1	0	0	0
Shrimps, slaters and crayfish	6	10	0	0	0	0
Mayflies	19	37	0	1	1	3
Stoneflies	7	27	0	1	0	0
Dragonflies	26	13	4	2	1	0
Water bugs	45	27	2	0	1	0
Water beetles	170	100	60	27	13	4
Alderflies and spongeflies	2	3	0	1	0	0
Caddis flies	71	95	3	7	1	4
Total number of species	398	368	72	41	21	13

Sources: The comparison is based on data derived from 156 sites in the National Pond Survey (unpublished data held by Pond Action) and 614 sites in the RIVPACS programme of the Institute of Freshwater Ecology (Wright *et al.*, 1996). The comparison is based on all invertebrate groups sampled in both surveys for which reliable published national distribution and status data are available. Further discussion of the methods used are given in Williams *et al.* (1998) and Wright *et al.* (1996). Numbers of taxa given by Wright *et al.* (1996) in their Table 1 were modified as follows to enable comparisons to be made: *Argulus foliaceus* was omitted from the Crustacea total; *Sigara (Sigara)* sp. was omitted from the Hemiptera total; water beetles in the family Scirtidae (4 taxa) were omitted from the Coleoptera total; Hydroptilidae (7 taxa) were omitted from the Trichoptera total.

Table 8.2 Number and area of standing water bodies in Britain in 1990.

Waterbody	number	%	Area (ha)	%
Ponds (up to 2.0 ha)[a]	297 300	97%	30 000	14%
Lakes (greater than 2.0 ha)	8900	3%	180 000	86%

[a] Assumes that 25% of ponds in the size range 1.0–5.0 ha in CS1990 were 2.0 ha or below. Sources: Barr *et al.* (1993), Barr *et al.* (1994.)

numerous in woodland and in the grassland dominated west of the country. Pond densities in arable dominated landscapes are lower (Williams *et al.*, 1998). Although enormous losses of ponds are thought to have occurred in Britain (in the order of 75% of those existing at the end of the nineteenth century), ponds are still common. Indeed, in Britain, the majority of standing waters are ponds; at least 97% of all discrete water bodies are less than 2 ha in area and even by area small water bodies probably represent some 15% of the total area of standing waters (Table 8.2).

The DETR Lowland Pond Survey has shown that, in Britain, ponds remain numerous in the landscape. However, this survey has also provided the first evidence of the very widespread impairment of ecological quality of ponds. Ponds in the lowland landscape of Britain were found to support on average only half the expected number of wetland plant species, when compared to minimally impaired ponds in similar landscape types. Aquatic (i.e. submerged and floating leaved) plant assemblages were particularly degraded, with the average pond in the countryside supporting only one-third the number of species expected. The Lowland Pond Survey also provided the first quantitative evidence of the extent to which non-native species have entered the British semi-natural flora; one in six wetland submerged plants recorded in the Lowland Pond Survey were non-native species, a local example of the world-wide phenomenon of erosion of biodiversity by alien species.

Hydrologically, ponds are very varied, and may be seasonal or permanent. Indeed, water permanence in ponds is one of the most important environmental variable shaping community types, with only pH-related variation being more significant (Pond Action, unpublished data). In national surveys of ponds, the primary grouping is into acid and base-rich sites; after this ponds are grouped according to whether they are permanent.

Perhaps the single most important hydrological feature of ponds (especially those fed mainly by surface runoff) is that many have very small catchments. This means that, unlike most rivers and lakes, many ponds in Britain have catchments wholly protected from the impacts of intensive rural and urban landuse, and the pollutants associated with intensive land use. Consequently it is possible to find ponds in non-intensively managed landscapes which are amongst the least impacted by human activities of all freshwater ecosystems. However, ponds that are impacted by pollutants are even more vulnerable than rivers and lakes because their small volumes provide little ability to dilute the impacts of pollutants.

Although it is quite simple to protect ponds from the effects of many degrading impacts derived from intensive land use (point source effluents and diffuse pollutants), many ponds are exposed to atmospheric deposition. In landscapes vulnerable to acidification, both upland and lowland, it is likely that very large numbers of small waterbodies are impaired by acidification. At present, there are few systematic data for small water bodies to assess this impact, although it seems likely that the same trends as those shown in the extensive and detailed studies of acid lakes will be apparent. No data are currently available on the effects of nutrient-enriched rainfall on the ecology of ponds.

8.2.2 *Management of ponds*

Lack of information about the ecology of ponds, and their popularity and familiarity, has led to the development of a wide variety of misconceptions about their ecology. This has been most apparent in the management of ponds for nature conservation, which has been largely dominated by the belief that maintaining open water by the physical removal of colonising vegetation and silt was the primary need for all ponds. In fact, the most important factor (as with all other freshwaters) is to ensure that water quality is maintained, usually by protecting the catchment, with physical, 'fine-tuning' management ponds in specific situations (Biggs *et al.*, 1994).

In Britain, the widespread neglect of ponds once used for watering farm livestock has led to the general assumption that ponds have no uses in the modern landscape. However, the DETR Lowland Pond Survey shows that a considerable proportion of ponds have modern amenity uses (for fishing and shooting for example, as well as providing visual diversity in the landscape), some of which are also compatible with protection of biodiversity. In addition, increasing numbers of ponds are being created to balance flood flows from urban areas, and to assist in 'source control' of pollutants.

The primary threats to small water bodies in Britain come from two main sources: pollution and in-filling (Williams *et al.*, 1998). Climate change also seems likely to pose a particular risk for seasonal ponds, which often support exceptional assemblages of specialised plant and animal species, many of which could completely disappear with slight shifts in climate (Collinson *et al.*, 1995). In pristine landscapes rich in natural water bodies, loss of seasonal ponds would probably be compensated for by permanent ponds becoming seasonal. In the fragmented British landscape, this process is not likely to operate effectively. For example, seasonal water bodies in the New Forest in southern England, which are exceptionally important freshwater ecosystems supporting a wide variety of rare and vulnerable species, could disappear.

8.2.3 *Monitoring of ponds*

The tendency to underestimate the importance of ponds has meant that there has been little monitoring of them. In Britain, the first national survey of pond quality has only recently been completed (Williams *et al.*, 1998). The authors are currently working on the establishment of a monitoring network building on the methods and sites of the National Pond Survey. Work is also currently in progress with the UK Environment Agency to develop a set of new monitoring techniques which will assess the quality of ponds against a minimally impaired baseline (Williams *et al.*, 1996, 1998). This system, the Predictive System for Multimetrics, combines the UK approach to river biological monitoring developed in the River Invertebrate Prediction and Classification Scheme programme (RIVPACS) with North American multimetric concepts of environmental monitoring (Wright *et al.*, 1993; Karr, 1991). First demonstrations of this technique have recently been made.

8.3   Acidification

Acidification is probably the most serious environmental problem facing the UK's upland lake ecosystems, yet the extent and severity of this problem has only recently been recognised. Gorham (1958) pointed out the threat of 'acid rain' to upland lakes but the first field evidence identifying the problem was not obtained until the late 1970s when a survey of the chemistry and biology of lakes in the Galloway region of Scotland showed the presence of highly acidic and, in some cases, fishless lakes (Wright and Henriksen, 1980, Harriman et al., 1987). Shortly afterwards Flower and Battarbee (1983), showed from diatom analysis of lake sediments, that many of these Galloway lakes had experienced declines in pH of between 0.5 and 1 pH unit over the last 100 years. They argued that the only plausible cause of such rapid and recent change was acid deposition.

In addition to the studies in Galloway, research in the Loch Ard forest region of central Scotland revealed significant differences in the acidification status of streams draining forested and moorland catchments (Harriman and Morrison, 1982). Streams draining catchments with a large proportion of conifer forest were frequently fishless and the authors concluded that the trees were increasing acidic deposition by intercepting pollutants more effectively than heathland and grassland vegetation. Similar studies in Wales (Stoner et al., 1984) confirmed the role of forests in increasing the acid deposition of acidic pollutants.

Despite these, and similar conclusions from studies in southern Scandanavia, other hypotheses were proposed, namely that the acidification was natural (e.g. Pennington, 1984) or the result of land-use changes (Rosenquist, 1978). A number of major projects, culminating in the Surface Waters Acidification Programme (SWAP) were designed to resolve the debate, resulting in a consensus that acid deposition was indeed the cause of the problem (Mason, 1990).

Since this realisation, there has been substantial progress in quantifying the impact of acidification on lake ecosystems throughout Britain. Sediment core work (Figure 8.2) has shown that the most intense acidification has occurred at sensitive sites in regions with the highest deposition of acidity (sulphur plus nitrogen), notably in north and central Wales, the Cumbrian Lake district, southwest Scotland, the Grampian region of Scotland and the eastern part of Northern Ireland (Battarbee et al., 1988). In northwest Scotland where acid deposition is low, the most sensitive lakes are affected, but only to a limited degree. In contrast, while much of the lowlands of England are subject to high levels of acid deposition, most surface waters are unaffected because of their high neutralising capacity. Exceptions to this are some of the regions of southwest and southern England (such as Dartmoor and the New Forest) where granites or well-weathered sandstones occur.

The sensitivity of a lake to acid deposition can be judged from its non-marine base cation content. In general, lakes with calcium concentrations

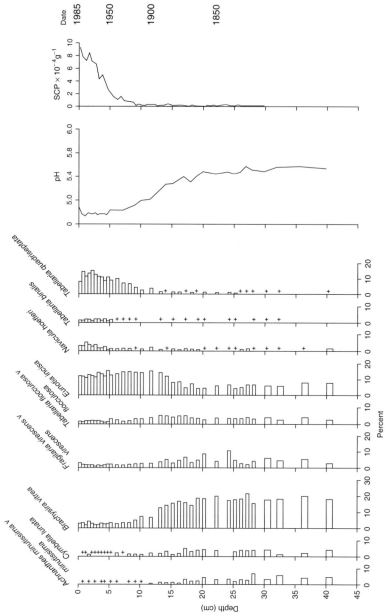

Figure 8.2 Summary diatom diagram, pH reconstruction and spheroidal carbonaceous particle (SCP) profile for a sediment core from the Round Loch of Glenhead, Galloway, Scotland. (From Jones *et al.*, 1989, and Birks *et al.*, 1990.)

greater than 200 µeq l⁻¹ are not sensitive to the levels of deposition experienced in the UK. For lakes with lower calcium levels the degree of acidification tends to be proportional to the sensitivity (expressed as calcium or total base cation levels) and to the loading of acidity (usually expressed as Keq $H^+ y^{-1}$).

As acidification has taken place at such sites, bicarbonate alkalinity has been reduced and lake-water pH has fallen, and this has been accompanied by an increase in dissolved aluminium concentrations. Laboratory and field experiments have shown the toxic effects of low pH, or low pH and elevated aluminium in combination, on aquatic biota in low calcium waters, explaining to a large degree the impoverishment of invertebrate faunas and the loss of salmonid fish populations in acidified lakes. Coloured waters with significant organic carbon tend to have a greater proportion of aluminium in a non-toxic complex form, and a recent study (Roy and Campbell, 1997) suggests that soluble organic matter reduces the toxicity of inorganic labile forms of aluminium. Most components of freshwater ecosystems, e.g. algae, aquatic macrophytes, zooplankton, benthic invertebrates and fish, are directly affected by increasing acidity, whereas some birds and amphibians are affected more indirectly (Morrison, 1989, Beebee et al., 1990, Ormerod et al., 1986).

Some amelioration of the problem has been achieved at individual sites, especially in Norway and Sweden, by catchment or lake liming. This is only considered to be a temporary palliative, and only a reduction in the emissions of sulphur and nitrogen can solve the problem at source. However, as both sulphur and nitrogen compounds can travel long distances in the atmosphere and across national boundaries, emission controls require international agreements. The long-term target is for emissions in Europe to be reduced so that acid deposition at the most sensitive sites is reduced to pre-acidification levels, which in theory, should allow a full recovery to take place. This reduced level of deposition is the 'critical load' for the site. For UK, critical load values have been calculated for surface waters across the country. In this exercise, the calculations were based on the results of water chemistry samples from lakes selected to be the most sensitive in each 10 km² of the country (Critical Loads Advisory Group (CLAG), 1995). Maps of critical loads have been produced for most European countries and, when these are compared with maps of acid deposition, they can be used to determine how much the deposition must be reduced at different sites or regions for recovery to take place. The critical load protocol thus targets emission reductions to provide the greatest possible recovery at the most sensitive sites.

Nevertheless, recovery may be slow. This is mainly because the release of accumulated sulphate in catchments is a slow process and also because some of the acidity in lakes is derived from nitrogen deposition. As catchment soils become progressively saturated by deposited nitrogen, nitrate leaching will occur and offset reductions in acidity expected from reductions in sulphur deposition. Dynamic models (Jenkins et al., 1997) show that substantial reductions in both sulphur and nitrogen are needed at many sites if critical loads are to be achieved. Biological recovery will probably be even slower than chemical recovery as the recolonisation of acidified lakes depends not only on water chemistry but also on the proximity and accessibility of plant and animal populations to the site and their speed of dispersal.

A network of acid waters monitoring sites was established in the UK in 1988 (Patrick *et al.*, 1991). Data from these sites are evaluated each year for changes in chemistry and biology (diatoms, invertebrates, aquatic macrophytes and fish). An equivalent deposition network was also established. After almost ten years, despite some reductions in sulphur deposition, the network so far shows few clear national trends. However, major reductions in sulphur emissions are planned over the next 10 years. These changes should be clearly identifiable at the deposition monitoring sites, and will provide an opportunity to assess how the chemistry and biology of lakes responds to known changes in acid deposition.

8.4 Eutrophication

8.4.1 *Introduction*

Accelerated eutrophication due to man's influence – rural and urban development – resulting in nutrient input rates much higher than those characteristic of former centuries is a major issue in UK lakes. The term *eutrophication* comes from the Greek *eu* meaning plenty, and *trophos* meaning food/feeding. Thus, a eutrophic lake is one liberally supplied with nutrients, i.e. nitrates, phosphates and silica. Major assessments and understanding of the causes and effects of eutrophication started in the early 1960s (e.g. Vollenweider, 1968; Millway 1970, OECD, 1982). The literature is thus enormous and reflects the world-wide concern over burgeoning eutrophication – and not just the process of inputting of nutrients *per se*, but more the manifestations of the enrichment. Of the many of these, the most noticeable are dense, durable algal blooms and related declines in macrophyte abundance and species diversity, leading to changes in other communities and trophic levels (e.g. Harper, 1986). A recent overview is that of Harper (1992), while Bailey-Watts (1994) has reviewed the situation for Scotland. A survey of eutrophication studies funded by the former Nature Conservancy Council (Bailey-Watts and May, 1991) also highlighted the widespread incidence of accelerated eutrophication in the UK. Excellent treatments on eutrophication and lake restoration are those of Sas (1989) and van Liere and Gulati (1992).

Major UK studies including attention to diffuse nutrient loadings and 'export coefficients' for land of different types, and for point-source inputs (e.g. sewage treatment works and textile factories using phosphorus-based chemicals) include Lough Neagh, Northern Ireland (e.g. Alexander and Stevens, 1976; Foy and Bailey-Watts, 1998; Foy *et al.*, 1982; Jordan and

Smith, 1985 and Smith, 1977; Stevens and Stewart, 1982) and Loch Leven (e.g. Bailey-Watts and Kirika, 1987, 1996; Holden 1975). See also Harper and Stewart (1987). More focused work includes that on nutrient enrichment from septic tanks (Patrick, 1988), overwintering geese (Hancock, 1982), forestry plantations (Harriman, 1978; Bailey-Watts, Kirika and Howell, 1988).

8.4.2   Nutrients

Of the three nutrients mentioned above, phosphorus (P) is most associated with lake eutrophication, in that in the temperate zone at least, it is the main nutrient limiting primary productivity of the phytoplankton (Reynolds, 1984). The other nutrients are still relevant however; indeed, the seasonal depletion of nitrate-nitrogen (NO_3-N) due to uptake by planktonic algae and denitrifying bacteria, leads to declines or a lack of increase in the abundance of all species except nitrogen-fixing blue-green algae (cyanobacteria) which augment their requirements by drawing on the dissolved elemental atmospheric nitrogen dissolved in the water (see e.g. Fay *et al.*, 1968). Depletion of silica (SiO_2) also affects production of diatoms, the only algae known so far with an absolute requirement for this nutrient in that they cannot even replicate DNA without it (Sullivan and Volcani, 1981). The wax and wane in silica availability can thus be considered as a major factor in lake phytoplankton ecology. Incidentally, very few silica loadings have been assessed for UK lakes other than Lough Neagh (e.g. Gibson, 1981) and Loch Leven (e.g. Bailey-Watts, Smith and Kirika, 1989).

8.4.3   Effects of eutrophication

Setting aside the influences of lake morphology and other physical factors on the chemical and biological effects of eutrophication (see below), there is an enormous literature on field and laboratory studies supporting the view that phosphorus enrichment results in increased primary production and biomass. By the same token, reductions in phosphorus loading eventually result in a decline in phytoplankton production (OECD, 1982).

The effects of phosphorus entering a lake on the amounts of phytoplankton produced in the first place, and on what is actually 'seen', per unit of phosphorus supply, depend very much on three interrelated features: bio-availability of the phosphorus (Reynolds, 1984), its origin (Dillon and Kirchner, 1974), and the timing/schedule of its entry to the system. Effluents from sewage treatment works and waste from fish and poultry farms are commonly rich in bio-available, soluble- or molybdate-reactive phosphorus (SRP/MRP), or at least in a form readily hydrolysable to SRP. Such sources are characteristically of low volume, and are largely independent of runoff. These 'point-sources' contrast markedly with the diffuse, comparatively 'lower grade', particulate phosphorus-dominated influxes that are governed by rainfall-runoff events and thus accompanied by large volumes of water. The contrast between the two sources in terms of water volume is of further significance. The potential to produce more algae, and see these manifested as elevated biomass (algal blooms, etc.), is considerably greater in the case of the point-sources. This is because of the higher bio-availability of the phosphorus and the relatively minor influence on hydraulic flushing which controls losses of planktonic organisms out of a lake.

Bailey-Watts (1994, 1998) and Bailey-Watts, May and Lyle (1992) identify a suite of 'sensitivity' factors (S) that determine – in conjunction with pressures (P′) to distinguish it from the nutrient on a lake by way of eutrophication – the responses (R) in terms of phytoplankton production, biomass and species composition i.e.

$$P' + S = R$$

Hydraulic flushing is a particularly important sensitivity factor in small lakes, but also – at times of thermal stratification, see section 8.1.1) – in large systems too. Other sensitivity factors are summarised in Table 8.3.

8.4.4   *Internal phosphorus loading*

Before turning attention to other issues central to understanding the causes and effects of lake eutrophication, the phenomenon of 'internal phosphorus loading' is worthy of mention. Releases of phosphate ions occur especially under conditions of low redox potential, in summer for example. Under warm calm conditions in shallow lakes, in particular where temperatures at the surface sediment may be much the same as the surface water), the phosphorus fluxes can compare with the external loadings (Bailey-Watts and Kirika, 1996). What is more, such releases are of the highest bio-available form, and do not affect flushing rates at all. This source can be viewed as a legacy of mis-use of lakes, and accelerated enrichment over many decades. Indeed, failures – or qualified successes at the most – to reverse eutrophication trends is commonly attributed to this phenomenon. Certainly, the conditions under which phosphorus releases occur also enhance the 'draw-down' of nitrate by de-nitrifying bacteria, and this – along with the grazing down of the smaller phytoplankton elements by filter-feeding Cladocera – 'opens the window' for the (usually large, gas-vacuolate) nitrogen-fixing cyanobacteria referred to above.

Observations on shallow systems – of which Loch Leven in Scotland and Lough Neagh in Northern Ireland are the best documented – suggest, however, that lakes are not uniform in their phosphorus release characteristics. At Loch Leven, it appears that massive releases of phosphorus are short-lived, particularly if warm calm weather conditions come to an end. Indeed,

Table 8.3 Some 'pressure' and 'sensitivity' factors enhancing the production and abundance of phytoplankton.

'Pressure' factors, mainly catchment characteristics	*'Sensitivity' factors, mainly in-loch features*
A rich, urbanised catchment – because this will export nutrients from relatively low volume – high concentration point sources (e.g. fish farms and sewage treatment works) which are often rich in SRP – the most immediately available form of P to algae and cyanobacteria (Bailey-Watts and Kirika, 1987); this is in contrast to relatively large volume-low concentration, diffuse runoff sources which are comparatively poor in SRP – and richer in less bio-available, particulate P (Bailey-Watts *et al.*, 1994, 1998)	Intrinsically clear, i.e. non-humic stained water – to maximise photosynthetic activity (Lyle and Bailey-Watts, 1993; Bailey-Watts *et al.*, 1992)
	Low flushing (long hydraulic residence time) – to minimise loss of cells through washout (Bailey-Watts *et al.*, 1990; Bailey-Watts *et al.*, 1992)
	Relatively shallow water e.g. mean depth of 3–5 m, or where overall depth much greater, thermal stratification (Lyle and Bailey-Watts 1993; Bailey-Watts *et al.*, 1992)
A low catchment-to-lake area (and volume) ratio and/or dry weather regime – to minimise rain-related runoff and thus flushing of the lake	High winter-time levels of major nutrients – to enhance phytoplankton production
	Rich ion content (high conductivity) to support nutrient sequestration
	Low grazing pressure – especially from *Daphnia* which has a preference for small algae (Bailey-Watts, 1982, 1986)
	Low susceptibility to parasitism by e.g. chytrid fungi (Bailey-Watts and Lund, 1973)

it is tempting to suggest that significant algal growth potential is only realised if a reasonable innoculum of planktonic algae already exists at the time of the releases. Phosphorus releases in this situation could be viewed therefore, as a 'red herring'. At Lough Neagh, however, equally prominent amounts of phosphorus are released from the sediments, but high concentrations are maintained for longer periods. Much of the literature concentrates on phosphorus releases from – and utilisation of nitrate by de-nitrifying bacteria at – the surfaces of deep-lying hypolimnia with reduced/anoxic muds. Similarly important releases occur in rich, shallow lakes where the sediment surface in summer is of much the same temperature as the surface water. However, phosphorus remobilisation can also occur in littoral sediments (Drake and Heaney, 1987).

8.4.5   *A note on 'blooms' the perception according to species composition*

Curiously, the most prominent 'blooms' which are of real concern in commonly imparting tastes and odours, interfering with leisure activities and being highly toxic, often do not constitute the annual biomass maxima. Peak, lakewide, biomass is more usually associated with populations composed of, for example, small centric diatoms which cloud the water but do not form clumps and scums in the manner of the 'classic' blue-green algae. Such maxima also occur primarily in winter or early spring, and as such are less noticed by the public. On the other hand, it is likely that the deposition of organic and inorganic material on to the sediments following the collapse of the diatoms constitutes an input of material, some of which may be mineralised and released back into the water column later in the year. In view of the particular importance of phytoplankton and the pelagic environment in lake ecology and issues over eutrophication and restoration, the reader is referred to the seminal books of Reynolds (1984, 1997).

8.4.6   *Restoring eutrophic lakes: prospects*

Reference to the suite of factors enhancing or suppressing phytoplankton production and biomass – and thus, molding much of the qualitative and quantitative features of the biota at higher trophic levels – should at the very least indicate clearly why waters such as the deep, humic-stained, nutrient-poor Loch Ness manifest low phytoplankton densities *per se*, and a low ratio of chlorophyll*a* content per unit phosphorus loading (Figure 8.3; Bailey-Watts, 1998) and the rich shallow Leven is the antithesis of this. Such data are important in assessing the relative sustainability of different waters to eutrophication. However, phytoplankton species (e.g. the diatom *Asterionella formosa*) indicative of at least mild eutrophication have been recorded in Loch Ness (e.g. Bailey-Watts and Duncan, 1981) and even very rich lakes can manifest clear water periods more akin to oligotrophic systems. In spite of the plethora of factors controlling the detailed phytoplankton species composition and 'succession' in lakes, good progress in simulating major seasonal changes in these features has been made. The on-going development of sophisticated dynamic models at the Institute of Freshwater Ecology (e.g. Hilton, Irish and Reynolds, 1992) represents an extremely valuable advance in our ability to predict the effects of eutrophication and assess the efficacy of measures to reverse enrichment trends. The value will increase, too, as weather forecasting improves. Equally, however, the simpler empirical models of, for example, Dillon and Rigler (1974), Kirchner and Dillon (1975), Vollenweider (1968, 1975 and 1976) retain their place on the basis of recognising (1) the importance of phosphorus as the main limiting nutrient in lakes in the UK and (2) the importance of gross physical features on the manner in which the nutrient is utilised.

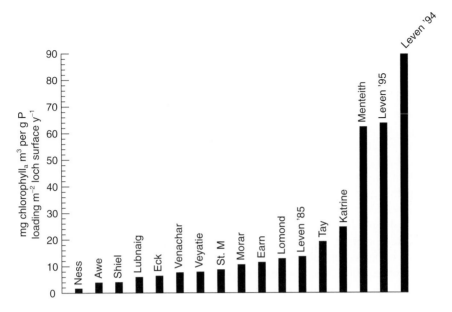

Figure 8.3 A ranking of 15 Scottish lochs according to the amounts of phytoplankton observed per unit of total phosphorus loading. (From Bailey-Watts, 1998.)

8.4.7 *The prospects of reversing eutrophication trends*

As part of the on-going studies on Loch Leven (now in their 31st year), Bailey-Watts, Gunn and Kirika (1993) listed some 30 eutrophication reversal strategies that have been executed with various degrees of success around the world. The schemes fall into two categories: those aimed at stemming nutrient inputs in the first place; and those carried out with the view to controlling existing algal blooms. Examples of the former are the implementation of phosphorus reduction facilities at sewage treatment works, the promotion of phosphorus-free detergents and the maintenance of 'buffer strips', while the latter includes harvesting algae, increasing water through-put and maintaining macrophyte stands. Trials on a number of these methods have been made at Loch Leven and Lough Neagh. The work on Loch Leven is of particular interest because any signs of improving water quality there would be viewed as a success, bearing in mind the very efficient manner in which this lake converts its phosphorus income into algae. There are some subtle signs of improvement, with recent total P levels remaining below the environmental quality standard (EQS) of 40 µg l^{-1} set by the local statutory authority for longer periods than in the past. In contrast, the corresponding EQS of 15 µg l^{-1} for chlorophyll*a* (which is the more concerning than phosphorus *per se*) is still rarely achieved. Bailey-Watts and Kirika (in press) show a net reduction of some 30% in the external loading of phosphorus to Loch

Leven between 1985 and 1995. However, a significant proportion of the decrease is due to the fact that 1995 much drier year than 1985.

8.5   Future changes in UK lakes and ponds

The extent to which the hydrology of the UK will change in future years is unknown. Nevertheless, whatever the future scenarios – but assuming a change of some sort in temperature, rainfall and evaporation – they will impact on all of the aspects summarised in this chapter. Almost certainly, as now, changes will not be uniform throughout the UK. Thus, in response to shifts in the main weather variables, lakes and ponds could change in depth and volume, and this would impact on chemical dynamics and biota. Changes in a considerable suite of factors and processes are likely. Changing physical and hydrographical regimes would affect, for example, water throughput rates, temperature, stratification and mixing patterns, and the rates and nature of fluxes of solutes and particulates between sediment and water. As a consequence, the seasonality, population densities, growth rates and metabolism, and possibly species diversity as a whole can be expected.

The effects of changing hydrology on acidification and eutrophication could be also considerable. As an example, raised temperatures could result in enhanced mineralisation of soil and litter components and increased runoff of these to standing waters. As shown from Loch Leven, shifts in rainfall will also affect the ratio of phosphorus loadings from point-sources (rich in the nutrient) and inputs from diffuse sources (comparatively low in phosphorus).

References

Alexander, G.C. and Stevens, R.J. (1976) Per capita phosphorus loading from domestic sewage. *Water Res.*, **10**, 757–64.

Bailey-Watts, A.E. (1982) The composition and abundance of phytoplankton in Loch Leven 1977–1979 and a comparison with the succession in earlier years. *Int. Revue Hydrobiol. Hydrogr.*, **67**, 1–25.

Bailey-Watts, A.E. (1986) Seasonal variation in phytoplankton size spectra in Loch Leven. *Hydrobiologia*, **33**, 25–42.

Bailey-Watts, A.E. (1994) Eutrophication, in Maitland, P.S., Boon, P.J. and McLusky, D.S. (eds), *The Fresh Waters of Scotland: A National Resource of International Significance*. Wiley, London. pp. 385–411.

Bailey-Watts, A.E. (1998) The phytoplankton ecology of the larger Scottish lochs. *Bot. J. Scotland*, **50**, 63–92.

Bailey-Watts, A.E. and Duncan, P. (1981) The ecology of Scotland's largest lochs: Lomond, Ness, Morar and Shiel 4. The phytoplankton. *Monographiae Biol.*, **44**, 91–118.

Bailey-Watts, A.E. and Kirika, A. (1987) A re-assessment of the phosphorus inputs to Loch Leven (Kinross, Scotland): rationale and overview of results on instantaneous loadings with special reference to runoff. *Trans. R. Soc. Edinburgh, Earth Sci.*, **78**, 351–67.

Bailey-Watts, A.E. and Kirika, A. (1996) A brief review of Loch Leven studies and some new findings on the sediments as an unusual diffuse source of phosphorus, in Petchey, A., D'Arcy, B. and Frost, A. (eds), *Diffuse Pollution and Agriculture*, Proceedings of the SAC-FRPB Conference held at the University of Edinburgh, 12–13 April 1995. Scottish Agricultural College, Aberdeen. pp. 224–32.

Bailey-Watts, A.E. and Kirika, A. (in press) Poor water quality in Loch Leven (Scotland) in 1995, in spite of reduced phosphorus loadings since 1985: the influences of catchment management and inter-annual weather variation. *Hydrobiologia*.

Bailey-Watts, A.E. and Lund, J.W.G. (1973) Observations on a diatom bloom in Loch Leven, Scotland. *Biol. J. Linn. Soc.*, 5, 235–53.

Bailey-Watts, A.E. and May, L. (1991) *A Review of Freshwater Eutrophication Studies Funded by the Nature Conservancy Council: Their Contribution to the Assessment, Control and Prevention of Enrichment Problems in the Future*. Final report to the Nature Conservancy Council.

Bailey-Watts, A.E., Gunn, I.D.M. and Kirika, A. (1993) *Loch Leven: Past and Current Water Quality and Options for Change*. Final Report to the Forth River Purification Board. Institute of Freshwater Ecology, Penicuik.

Bailey-Watts, A.E., Kirika, A. and Howell, D.H. (1988) *The potential Effects of Phosphate Runoff From Fertilised Forestry Plantations on Reservoir Phytoplankton: Literature Review and Enrichment Experiments*. Final report to the Water Research Centre. Water Research Centre, Medmenham.

Bailey-Watts, A.E., Kirika, A., May, L. and Jones, D.H. (1990) Changes in phytoplankton over various time scales in a shallow eutrophic lake: the Loch Leven experience with special reference to the influence of flushing rate. *Freshwater Biol.*, 23, 85–111.

Bailey-Watts, A.E., May, L., Kirika, A. and Lyle, A.A. (1992) *Eutrophication case studies: phase II, an assessment based on desk analysis of catchments and summer limnological reconnaissances. Volume 1. An analysis of the whole spectrum of waters studied*. Final report to the Nature Conservancy Council for Scotland.

Bailey-Watts, A.E., May, L. and Lyle, A.A. (1992) *Factors Determining the Response of Waters to Enhanced Nutrient Enrichment*. Report to the Water Research Centre. Water Research Centre, Medmenham.

Bailey-Watts, A.E., Smith, I.R. and Kirika, A. (1989) The dynamics of silica in a shallow, diatom-rich Scottish loch I: stream inputs of the dissolved nutrient. *Diatom Res.*, 4, 171–90.

Barr, C.J., Bunce, R.G.H., Clarke, R.T. *et al.* (1993) *Countryside Survey 1990 Main Report*. Countryside 1990 Series, Vol. 2. Department of the Environment, London.

Barr, C.J., Howard, D.C. and Benefield, C.B. (1994) *Countryside Survey 1990. Inland Water Bodies*. Countryside 1990 Series, Vol. 6. Department of the Environment, London.

Battarbee, R.W., Anderson, N.J., Appleby, P.G. *et al.* (1988) *Lake acidification in the United Kingdom 1800–1986: Evidence From Analysis of Lake Sediments*. Ensis, London.

Beebee, T.J.C., Flower, R.J., Stevenson, A.C. *et al.* (1990) Decline of the natterjack toad Bufo calamita in Britain: palaeoecological, documentary and experimental evidence for breeding site acidification. *Biol. Conserv.*, 53, 1–20.

Biggs, J., Corfield, A., Walker, D., Whitfield, M. and Williams, P. (1994) New approaches to the management of ponds. *British Wildlife*, 5, 273–87.

Biggs, J., Williams, P., Corfield, A. *et al.* (1996) *Pond Survey 1996. Stage 1, Scoping Study.* Pond Action, Oxford and ITE, Merlewood.

Birks, H.J.B., Line, J.M., Juggins, S., Stevenson, A.C. and ter Braak, C.J.F. (1990) Diatoms and pH reconstruction. *Phil. Trans. R. Soc., London,* **B** 327, 263–78.

Central Office of Information (1974) *Britain 1974. An Official Handbook.* HMSO, London.

CLAG (1995) *Critical Loads of Acid Deposition for United Kingdom Freshwaters.* Critical Loads Advisory Group sub-group report on freshwaters. Institute of Terrestrial Ecology, Penicuik.

Collinson, N.H., Biggs, J., Corfield, A. *et al.* (1995) Temporary and permanent ponds: an assessment of the effects of drying out on the conservation value of aquatic macroinvertebrate communities. *Biol. Conserv.,* 74, 125–34.

Dillon, P.J. and Kirchner, W.B. (1974) The effects of geology and land use on the export of phosphorus from watersheds. *Water Res.,* 9, 135–48.

Dillon, P.J. and Rigler, F.H. (1974) The phosphorus-chlorophyll relationship in lakes. *Limnol. Oceanogr.,* 19, 767–73.

Drake, J.C. and Heaney, S.I. (1987) Occurrence of phosphorus and its potential remobilization in the littoral sediments of a productive English lake. *Freshwater Biol.,* 17, 513–23.

Fay, P., Stewart, W.D.P., Walsby, A.E. and Fogg, G.E. (1968) Is the heterocyst the site of nitrogen fixation in blue-green algae? *Nature,* 220, 810–12.

Flower, R.J. and Battarbee, R.W. (1983) Diatom evidence for recent acidification of two Scottish lochs. *Nature,* 20, 130–33.

Foy, R.H. and Bailey-Watts, A.E. (1998) Observations on the spatial and temporal variation in the phosphorus status of lakes in the British Isles. *Soil Use Manag.,* 14, 131–138.

Foy, R.H., Smith, R.V., Stevens, R.J. and Stewart, D.A. (1982) Identification of factors controlling nitrogen and phosphorus loadings to Lough Neagh. *J. Environ. Manag.,* 15, 109–129.

Gibson, C.E. (1981) Silica budgets and the ecology of planktonic diatoms in an unstratified lake (Lough Neagh, N. Ireland). *Int. Revue der gesamten Hydrobiologie und Hydrographie,* 66, 641–64.

Golterman, H.L. (1975) *Physical Limnology.* Elsevier, Amsterdam.

Gorham, E. (1958) The influence and importance of daily weather conditions in the supply of chloride, sulphate and other ions to freshwaters from atmospheric precipitation. *Phil. Trans. R. Soc., London,* **B** 241, 147–78.

Gorham, E. and Boyce, F.M. (1989) Influences of lake surface area and depth upon thermal stratification and the depth of the summer thermocline. *J. Great Lakes Res.,* 15, 233–245.

Hancock, C.G. (1982) Sources and utilisation of nutrients in the Loch of Strathbeg, Aberdeenshire. Unpublished doctoral dissertation, University of Aberdeen, Aberdeen.

Harper, D.M. (1986) The effects of artificial enrichment upon the planktonic and benthic communities in a mesotrophic to hypertrophic loch series in lowland Scotland. *Hydrobiologia,* 137, 9–19.

Harper, D.M. (1992) *Eutrophication of Freshwaters: Principles, Problems and Restoration.* Chapman and Hall, London.

Harper, D.M. and Stewart, W.D.P. (1987) The effects of land use upon water chemistry, particularly nutrient enrichment, in shallow lowland lakes: comparative studies of three lochs in Scotland. *Hydrobiologia*, 148, 211–29.

Harriman, R. (1978) Nutrient leaching from fertilised forest watersheds in Scotland. *J. Appl. Ecol.*, 15, 933–42.

Harriman, R. and Morrison, B.R.S. (1982) The ecology of streams draining forested and non-forested catchments in an area of central Scotland subject to acid precipitation. *Hydrobiologia*, 88, 251–63.

Harriman, R., Morrison, B.R.S., Caines, L.A., Collen, P. and Watt, A.W. (1987) Long term changes in fish populations of acid streams and lochs in Galloway, south-west Scotland. *Water, Air Soil Pollut.*, 32, 89–112.

Hilton, J., Irish, A.E. and Reynolds, C.S. (1992) Active reservoir management: a model solution, in Sutcliffe, D.W. and Jones, J.G. (eds), *Eutrophication: Research and Application to Water Supply*. Freshwater Biological Association, Ambleside. pp. 185–96.

Holden, A.V. (1975) The relative importance of agricultural fertilisers as a source of nitrogen and phosphorus in Loch Leven. *Ministry of Agric., Fisheries Food Tech. Bull.*, 32, 306–14.

Hutchinson, G.E. (1957) *A Treatise on Limnology*. Wiley, New York.

Jenkins, A., Renshaw, M., Helliwell, R. *et al.* (1997). Modelling surface water acidification in the UK. *Institute of Hydrology Report No. 131*. Institute of Hydrology, Wallingford.

Jones, V.J., Stevenson, A.C. and Battarbee, R.W. (1989) Acidification of lakes in Galloway, south west Scotland: a diatom and pollen study of the post-glacial history of the Round Loch of Glenhead. *J. Ecol.*, 77, 1–23.

Jordan, C. and Smith, R.V. (1985) Factors affecting leaching of nutrients from an intensively managed grassland in County Antrim, Northern Ireland. *J. Environ. Manag.*, 20, 1–15.

Karr, J. (1991) Biological integrity: a long-neglected aspect of water resource management. *Ecol. Appl.*, 1, 66–84.

Kirchner, W.B. and Dillon, P.J. (1975) An empirical method of estimating the retention of phosphorus in lakes. *Water Resources Res.*, 11, 182–183.

Lyle, A.A. (1981) Ten years of ice records for Loch Leven, Kinross. *Weather*, 36, 116–25.

Lyle, A.A. (1987) The bathymetry and hydrology of some lochs vulnerable to acid deposition in Scotland, in Maitland, P.S., Lyle, A.A. and Campbell, R.N.B. (eds), *Acidification and Fish in Scottish Lochs*. Institute of Terrestrial Ecology, Cambridge. pp. 23–34.

Lyle, A.A. and Bailey-Watts, A.E. (1993) *I. Effects of light attenuation by humic colouring and turbidity on chlorophyll production. II. Factors controlling lake stratification. Contributions to the Scotland and Northern Ireland Forum for Environmental Research (SNIFFER) programme on Eutrophication Risk Assessment*. Report to the Water Research Centre.

Lyle, A.A. and Smith, I.R. (1994) Standing Waters, in Maitland, P.S., Boon, P.J. and McLusky, D.S. (eds), *The Fresh Waters of Scotland*. Wiley, Chichester. pp. 35–50.

Mason, B.J. (ed.) (1990) *The Surface Waters Acidification Programme*. Cambridge University Press, Cambridge.

Millway, C.P. (1970) *Eutrophication in Large Lakes and Impoundments.* Report on Uppsala Symposium, May 1968. Organisation for Economic Cooperation and Development, Paris.

Morrison, B.R.S. (1989) Freshwater life in acid streams and lochs, in *Acidification in Scotland.* Scottish Development Department, Edinburgh. pp. 82–9.

OECD (1982) *Eutrophication of Waters, Monitoring Assessment and Control.* Organisation for Economic Cooperation and Development, Paris.

Ormerod, S.J., Allinson, N., Hudson, D. and Tyler, S.J. (1986) The distribution of breeding dippers (*Cinclus cinclus* (L.) Aves) in relation to stream acidity in upland Wales. *Freshwater Biol.,* 16, 501–7.

Patrick, S.T. (1988) Septic tanks as sources of phosphorus to Lough Erne, Ireland. *J. Environ. Manag.,* 26, 239–48.

Patrick, S.T., Waters, D., Juggins, S. and Jenkins, A. (eds) (1991) *The United Kingdom Acid Waters Monitoring Network: Site Descriptions and Methodology Report.* Ensis, London.

Pennington, W. (1984) Long-term natural acidification of upland sites in Cumbria: evidence from post-glacial lake sediments. *Freshwater Biological Association Annual Report.* 52, pp. 28–46.

Reynolds, C.S. (1984) *The Ecology of the Freshwater Phytoplankton.* Cambridge University Press, Cambridge.

Reynolds, C.S. (1997) Vegetation processes in the Pelagic: a model for ecosystem theory, in Kinne, O. (ed.), *Excellence in Ecology, 9.* Ecology Institute, D-21385, Oldendorf/Luhe, Germany.

Rich, D., Samson, W.A., Scott, R., White, J. and Whitfield, M. (1998) *Lowland Pond Survey 1996.* Department of the Environment, Transport and the Regions, London.

Rosenqvist, I.T. (1978). Alternative sources for acidification of river water in Norway. *Sci. Total Environ.,* 10, 39–49.

Roy, R.L. and Campbell, P.G.C. (1997) Decreased toxicity of Al to juvenile Atlantic salmon (Salmo salar) in acidic soft water containing natural organic matter: a test of the free ion model. *Environ. Toxicol. Chem.,* 16, 1962–9.

Sas, H. (1989) *Lake Restoration by Reduction of Nutrient Loading: Expectations, Experiences, Extrapolations.* Academia-Verlag Richarz, St Augustin.

Smith, I.R. and Lyle, A.A. (1979) *Distribution of Freshwaters in Great Britain.* Institute of Terrestrial Ecology, Cambridge.

Smith, I.R., Lyle, A.A. and Rosie, A.J. (1981) Comparative physical limnology, in Maitland, P.S. (ed.), *The Ecology of Scotland's Largest Lochs: Lomond, Awe, Ness, Morar and Shiel.*

Smith, R.V. (1977) Domestic and agricultural contributions to the inputs of phosphorus and nitrogen to Lough Neagh. *Water Res.,* 11, 453–59.

Smith, S.J., Wolfe-Murphy, S.A., Enlander, I. and Gibson, C.E. (1991) *The Lakes of Northern Ireland: An Annotated Inventory.* Countryside and Wildlife Research Series No 3. HMSO, Belfast.

Stevens, R.J. and Stewart, D.A. (1982) The effects of sampling interval and method of calculation on the accuracy of estimation of nitrogen and phosphorus loads in drainage water from two different size catchment areas. *Record. Agric. Res.,* 29, 29–38.

Stoner, J.H., Gee, A.S. and Wade, K.R. (1984) The effects of acidification on the ecology of streams in the upper Tywi catchment in West Wales. *Environ Pollut.*, 35, 125–57.

Sullivan, C.W. and Volcani, B.E. (1981) Silicon in the cellular metabolism of diatoms, in Simpson, T.L. and Volcani, B.E. (eds), *Silicon and Siliceous Structures in Biological Systems*. Springer-Verlag, New York, pp. 14–42.

Van Liere, L. and Gulati, R.D. (1992) *Restoration and Recovery of Shallow Lake Ecosystems in the Netherlands*. Proceedings of a Conference held in Amsterdam, The Netherlands, 18–19 April 1991. Kluwer, Dordrecht.

Vollenweider, R.A. (1968) *Water Management Research; Scientific Fundamentals of the Eutrophication of Lakes and Flowing Waters, with Particular Reference to Nitrogen and Phosphorus as Factors in Eutrophication*. Technical report DAS/CSI/68.27. Organisation for Economic Cooperation and Development, Paris.

Vollenweider, R.A. (1975) Input-output models with special reference to the phosphorus loading concept in limnology. *Schw. Z. Hydrobiologie*, 37, 53–84.

Vollenweider, R.A. (1976) Advances in defining critical loading levels for phosphorus in lake eutrophication. *Memorie dell' Istituto Italiano di Idrobiologia*, 33, 53–83.

Williams, P., Biggs, J., Dodds, L. *et al.* (1996) *Biological Techniques of Still Water Quality Assessment: Phase 1 Scoping Study*. R & D Technical Report E7. Environment Agency, Bristol.

Williams, P.J., Biggs, M., Whitfield, A., Corfield, G., Fox, D.C. and Adare, K. (1998) *Biological Techniques of Still Water Quality Assessment. Phase 2 Method Development*. R&D Technical Report E56. Environment Agency, Bristol.

Wright, R.F. and Henriksen, A. (1980) *Regional Survey of Lakes and Streams in SW Scotland*. SNSF Report IR72/80. Norwegian Institute for Water Research, Oslo.

Wright, J.F., Furse, M.T. and Armitage, P.D. (1993) RIVPACS – a technique for evaluating the biological quality of rivers in the UK. *Eur. Water Pollut. Control.*, 3, 15–25.

Wright, J.F., Blackburn, J.H., Gunn, R.J.M., Furse, M.T., Armitage, P.D., Winder, J.M., Symes, K.L. and Moss, D. (1996) Macro-invertebrate frequency data for the RIVPACS III sites in Great Britain and their use in conservation evaluation. *Aquatic Conservation: Marine and Freshwater Ecosystems*, 6, 141–67.

9

WETLANDS

Mike Acreman and Paul José

This chapter is concerned with wetlands, their role in the UK hydrological cycle and the factors which have led to varying levels of loss, protection and restoration. It covers both the physical hydrology aspects and the social, legal and political issues which affect the hydrology of wetlands indirectly.

9.1   What are wetlands?

Wetlands are a fundamental part of the UK's landscape, embracing a diverse range of habitats including marshes, fens (Plate 9.1), bogs, wet grasslands, carrs, floodplains and mudflats. They occupy the transitional zones between permanently wet and generally drier areas. They share characteristics of both environments yet cannot be classified unambiguously as either aquatic or terrestrial. It is the presence of water for some significant period of time which creates the soils, its micro-organisms and the plant and animal communities, such that the land functions in a different way from either aquatic or dry habitats. Indeed, Mitsch and Gosselink (1993) highlighted the fact that 'hydrology is probably the single most important determinant for the establishment and maintenance of specific types of wetlands and wetland processes . . . when hydrologic conditions in wetlands change even slightly, the biota may respond with massive changes in species richness and ecosystem productivity'. The Convention on Wetlands of International Importance, established in Ramsar, Iran, in 1972 (Davis, 1994), and signed by the UK in 1979, adopts an extremely broad approach and defines wetlands as:

> 'areas of marsh, fen, peatland or water, whether natural or artificial, permanent or temporary, with water that is static or flowing, fresh, brackish or salt, including areas of marine water, the depth of which at low tide does not exceed 6 m'.

The distribution within the UK of key wetlands classified as wet grasslands (which include floodplains, water meadows and coastal grazing marsh),

Plate 9.1 Wicken Fen, Cambridgeshire (Mike Acreman).

reedbeds and bogs (including raised bogs, intermediate bogs and blanket bogs) is shown in Figure 9.1.

9.2   Why are wetlands important?

In recent years there has been increasing awareness of the many valuable hydrological functions that wetlands provide free of charge (e.g. flood alleviation, groundwater recharge, retention of pollutants), as well as useful products (e.g. fish, timber), and attributes (e.g. biodiversity) as indicated in Table 9.1. Wildlife support is a particularly important aspect of UK wetlands. Over 3500 species of invertebrates, 150 species of aquatic plant, 22 species of duck and 33 species of wader have been identified living in UK wetlands, whilst all six of our native species of amphibian depend on wetlands for breeding (Merritt, 1994). Benstead *et al.* (1997) list key plants and birds

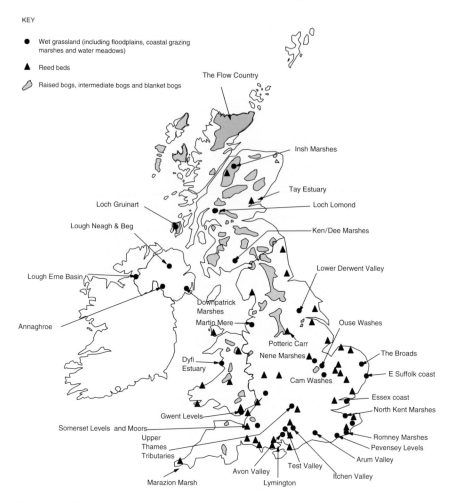

KEY

● Wet grassland (including floodplains, coastal grazing marshes and water meadows)

▲ Reed beds

Raised bogs, intermediate bogs and blanket bogs

The Flow Country

Insh Marshes

Tay Estuary

Loch Gruinart

Loch Lomond

Lough Neagh & Beg

Ken/Dee Marshes

Lower Derwent Valley

Lough Erne Basin

Downpatrick Marshes

Annaghroe

Martin Mere

Ouse Washes

Potteric Carr

Nene Marshes

The Broads

Dyfi Estuary

E Suffolk coast

Cam Washes

Essex coast

North Kent Marshes

Gwent Levels

Somerset Levels and Moors

Romney Marshes

Upper Thames Tributaries

Pevensey Levels

Arum Valley

Avon Valley

Test Valley

Itchen Valley

Marazion Marsh

Lymington

Figure 9.1 Some key wet grasslands (after Benstead *et al.*, 1997), reedbeds (after Hawke and José, 1996) and bogs (after Stoneman and Brooks, 1997) in the UK.

(Table 9.1) and amphibians, reptiles, invertebrates and mammals supported by wet grassland wetlands.

Wetlands are also important for archaeology; for example, the Sweet Track – the oldest built roadway in the world (3806–3791 BC) – was preserved in the Somerset Levels (Coles, 1995). In addition, the movement of water within and through the soils and vegetation of wetlands can influence on hydrological cycle in a way that can be beneficial to human populations. Such hydrological functions (Table 9.2) may include flood alleviation (Box 9.1) and water quality improvement (Box 9.2).

Table 9.1 Birds and plants of conservation concern on wet grasslands
(after Benstead *et al.*, 1997).

Plants	Birds
Red Data Book	Red list species (high conservation concern)
Downy-fruited sedge, tuberous thistle, greater yellow-rattle, creeping marshwort	Winter: hen harrier, merlin, twite Summer: quail, corncrake, black-tailed godwit All year: grey partridge, skylark, tree sparrow, linnet
Nationally scarce	Amber list (medium conservation concern)
Marsh-mallow, slender hare's ear, divided sedge, elongated sedge, narrow-leaved water-dropwort sea clover, snake's head fritillary, drawf mouse-ear, sulphur clover, sea barley, milk-parsley, wavy St John's-wort, cambridge milk parsley	Winter: Bewick's swan, whooper swan, bean goose, pink-footed goose, white fronted goose, barnacle goose, Brent goose, wigeon, shoveler, gadwall, pintall, peregrine, golden plover, common gull, short-eared owl, fieldfare, redwing, starling. Summer: garganey, pintail, spotted crake, ruff oystercatcher, whimbrel (on passage), curfew, redshank All year: teal, shoveler, pochard, kestrel, lapwing, snipe, barn owl, kingfisher, goldfinch

Table 9.2 Wetland products, functions and attributes.

Components/products	Functions/services	Diversity/attributes
Reeds Peat Hay Wildlife e.g. Natterjack toad, fen raft spider Fish Water supply	Groundwater discharge/ recharge Flood and flow control Shoreline/bank stabilisation Sediment retention Nutrient retention Recreation/tourism Water transport Archaeological preservation	Biological diversity Uniqueness/cultural heritage

Barbier *et al.* (1997) have produced guidelines for the economic valuation of wetlands which demonstrate the high value of wetland functions. Several studies have been conducted in the UK. For example, Hanley and Craig (1991) conducted a partial valuation of alternative uses of peat bog in northern Scotland's 'Flow Country'. They calculated the net present value of conserving the area as £327 per ha (in 1990 prices). In contrast, the net benefits of

Box 9.1 Flood attenuation by wetlands

The floodplain of the River Wye between Erwood and Belmont demonstrates clearly the flood attenuation effect of wetlands and was chosen as very suitable to test models of flood attenuation in the UK Flood Studies Report (NERC, 1975). The flow measurement stations at Erwood and Belmont have good quality flow data, even a high flows, and there are more than thirty years of record, containing floods of a range of magnitudes. The reach is 69.8 km long and has no important tributaries, thus the lateral inflow along the reach is small (some $14 \text{ m}^3 \text{s}^{-1}$) in comparison with the average annual flood discharge at Belmont ($560 \text{ m}^3 \text{s}^{-1}$). The reach contains large floodplains, 28.6 km², primarily between Witney and Bredwardine. During high flows (greater than about $400 \text{ m}^3 \text{s}^{-1}$ at Belmont) flood water is stored on the floodplain which plays a crucial role in reducing peak discharge from Erwood by up to 45% at Belmont (Figure 9.2) so giving important protection to the city of Hereford downstream.

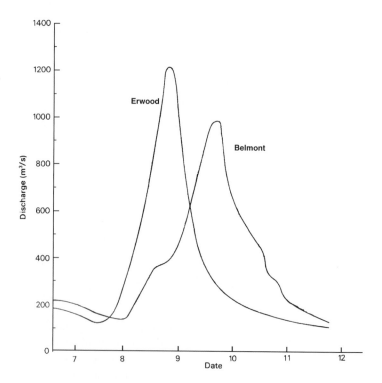

Figure 9.2 River Wye: flood of December 1960. (After NERC, 1975.)

Box 9.2 Nutrient removal by wetlands

Wetlands can play an important role in removing soluble phosphorus and nitrate from agricultural runoff which could cause eutrophication of receiving water bodies. Consequently, they can potentially be used as buffer zones in the management of areas vulnerable to nitrate pollution (Haycock and Burt, 1993). As part of the functional analysis of European wetland ecosystems (FAEWE; Maltby *et al.*, 1996), studies were undertaken of nutrient dynamics on river marginal wetlands in the catchment of the River Torridge in north Devon. Kismeldon Meadows is an SSSI, consisting of traditionally managed grassland on a gentle slope and narrow floodplain which are situated below improved managed grassland. Nutrient rich runoff flows into the site which contains several 'flush' zones, before entering the river. The soluble phosphorus load in the ditch draining the grassland was some 100 g over two days. This was converted into predicted phosphorous concentrations (Figure 9.3). If this loading reached the River Torridge unaltered, it would represent a risk to river water quality. At a measuring weir below the flushes, the observed concentrations were on average 73% lower than predicted, with 50% representing the smallest reduction (Russel and Maltby, 1995). Overall soluble phosphorus concentrations are a function of discharge. Overland flow reduces the role of the flush site in reducing phosphorus concentrations.

Baker and Maltby (1995) found that 95% of the nitrate (up to 4.5 mg l^{-1}) discharging into the same wetland was not leaving the system. This suggests that the nitrate is being stored or denitrified. The rate of denitrification was estimated at 3.5 mg $m^{-2}hr^{-1}$.

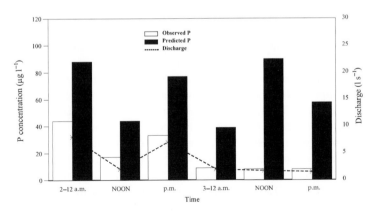

Figure 9.3 Observed and predicted phosphorus concentrations. (After Russel and Maltby, 1995.)

tree planting was negative, at £895 per ha, suggesting that it is only as a result of government incentive payments that planting has occurred. These payments have since been withdrawn. In another study, Bateman *et al.* (1995) estimated the recreation and amenity value of the Broads in East Anglia. They elicited an aggregate willingness-to-pay response to maintain the wetland area in its present condition of £32.5 million for households within a zone close to the wetlands and £7.3 million from elsewhere in UK.

These environmental services have been highlighted by many groups, such as the European Commission (1995), and particularly to support the worldwide wetland conservation effort (see, for example, Dugan 1990, which has been translated into nine languages). Because of their hydrological and chemical functions, wetlands have been described both as 'the kidneys of the landscape' (Mitsch and Gosselink, 1995). However, not all wetlands perform these functions to the same extent, if at all. Indeed, some wetlands perform hydrological functions which may be contrary to human needs, e.g. riparian wetlands may act as runoff generating areas thus increasing flood risk downstream.

Clearly not all wetlands perform all these functions; for example, reedbeds are particularly effective at removing nutrients, whilst floodplains are most effective at reducing flood peaks. For these key functions, there is well-documented scientific evidence. However, the extent to which other wetlands perform other functions is less certain. Furthermore, some wetlands perform functions which may be considered to be unhelpful to mankind. For example, some headwater river margins are zones of runoff generation and can augment floods.

9.3   Threats and loss

Despite the valuable products, functions and attributes exhibited by wetlands that are now becoming more widely appreciated, there is ignorance of these positive characteristics. For many people, wetlands conjure up a swamp full of slimy creatures, harbouring diseases, such as malaria. Indeed it is this view of wetlands as 'wastelands' that has led to extensive drainage and conversion of wetlands for intensive arable agricultural, and industrial or residential land. Thus, despite their importance, many UK wetlands are being lost, either temporarily or permanently, for a number of reasons including groundwater abstraction, drainage, separation of the river and its floodplain by embankment and lack of appropriate management (Acreman and Adams, 1998). Some figures for wetland loss are available. For example, English Nature (1997) suggests that since 1930, 64% of the wet grassland in the Thames valley, 48% in Romney Marsh and 37% in the Norfolk/Suffolk Broads has been lost. Some figures for coastal grassland loss are shown in Figure 9.4, but there is no national inventory from which to construct a

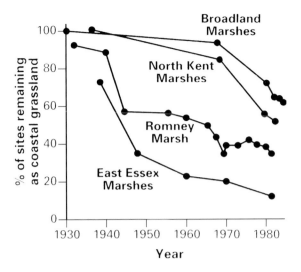

Figure 9.4 Decline in the extent of coastal grasslands in four areas of southeast England. (From Benstead *et al.*, 1997.)

comprehensive picture. Although a discussion document – *The NRA's Role in Wetland Conservation* – was prepared by the NRA (1995a), no definition or classification scheme and no national wetlands strategy has been agreed, although the Environment Agency has established a Wetland Liaison Group to address some of these issues.

In a desk study, English Nature (1996) assessed the perceived impact of abstraction at 137 wetland SSSIs and concluded that 15 were at high risk and a further 31 were at medium risk. In the high-risk sites, public water supply abstraction was implicated at seven sites with agricultural use being the other possible main threat. The English Nature study was based on hydrogeological assessments, as ecological impact studies did not exist. Extending this work, the report entitled *High and Dry* (Biodiversity Challenge, 1996) claimed that abstraction is known, or thought, to affect currently 151 wetland sites throughout England and Wales with a further 100 wetlands at future risk. It was suggested that the scale of the problem could be even greater if non-statutory sites were to be included.

The Environment Agency did not accept the 'future risk' category referred to in the English Nature and Biodiversity Challenge reports. The Agency and English Nature are jointly reviewing information on the sites from these two reports and others that have been identified subsequently (over 350 in total). The work is not yet complete, but at some sites the ecological impact has been found to be minimal, even where relatively large changes in groundwater levels have occurred. In contrast, at some sites

---

Box 9.3 **Example wetlands affected by abstraction from the third periodic review** (Environment Agency, 1998)

Site category	Example sites	Code
Habitat Directive sites (put forward in England by English Nature)	Yare Broads and marshes, Norfolk	1
	Derwent Ings, Yorkshire	2
	Coston Fen, Norfolk	3
SSSIs (put forward in England by English Nature)	Woodwalton Fen, Cambridgeshire	1
	Taw Marsh, Devon	2
	Pevensey Levels, Sussex	3
Non-SSSI sites affected by excessive abstraction, but not exceeding licensed quantity (put forward for England and Wales by the Environment Agency)	Burntwood Pools, Staffordshire	1
	Aqualate Mere, Shropshire	2

Code: 1, Evident problem as a result of water company abstraction; 2, Evident problem of drying out, but relationship to water company abstraction currently unclear; 3, Concern about impact of abstraction, but wildlife features not currently at high risk – further investigations agreed between English Nature/Country Commission for Wales and the Environment Agency. If these show a serious impact, a scheme may subsequently be prepared for implementation.

---

where only minor changes in groundwater have been recorded, the impact on the ecology has been significant, reflecting the sensitivity of certain species to changes in the hydrological regime. Preliminary results of the study suggest that significant problems related to groundwater abstraction are apparent at 20 sites. Ecological problems have been identified at around 60 other sites, but the cause is not certain. At 96 further sites, abstraction for public water supply or irrigation is an issue, but the impacts are not significant.

As part of the periodic review of water resources, water companies are producing their third set of Asset Management Plans (AMP3) that contain strategies for water efficiency and conservation. The AMP3 process includes identification of a list of 51 wetland SSSIs (mostly fens and bogs) affected by water abstraction. Water companies have been asked to cost remedial schemes at 21 of them and investigations at the remainder. Examples are given in Box 9.3.

Figure 9.5 UK Ramsar sites, September 1998.

9.4 Wetland protection

Various laws, conventions and policies have been established to conserve wetlands – for example, the UK signed the Ramsar Convention in 1990 – but their effectiveness has been limited. For example, Neild and Rice (1996) carried out a review of the UK compliance with the Ramsar Convention and concluded that although 85 sites had been designated (now 124, Figure 9.5), 80 were still on the candidate list. Over half of the designated sites had suffered recorded damage between 1993 and 1996 or remain under threat. Many should be placed on the Montreux Record (the Conventions list of sites where changes in ecological character have occurred, are occurring or are likely to occur). Neild and Rice concluded that part of the problem is that existing domestic legislation is too weak to implement the Convention and requires strengthening.

9.5 Wetland management objectives

The UK was one of 157 countries which signed the Convention on Biological Diversity in June 1992. This required the development of national plans or programmes for the conservation and sustainable use of biological diversity. In January 1994, the UK Government published *Biodiversity: The*

Box 9.4 **Examples of objectives**

	Overall management objective	Water level objective	Approach used
Pevensey Levels wetland	Restore and maintain ecology at 1970 levels	Maintain ditch water levels not more than 300 mm below ground level Mar.–Sept. not more than 600 mm below ground level Oct.–Feb.	Expert opinion research on water requirements of ecology of wetland species
Somerset Moors and Levels	Restore numbers of breeding waders to 1970 level	Raise water levels in Winter to produce splash-flooding and maintain water levels within 200 mm of ground surface in Spring	Expert opinion on ecology of wading birds
Chippenham, Wicken, Fulbourn Fens	Protection of notifiable vegetation communities	Target flows identified in the River Granta and Lodes	Lodes-Granta groundwater model, test pumping, hydrological studies

UK Action Plan and set up the UK Biodiversity Steering Group with a broad cross-sectoral membership. The Steering Group's report, endorsed by Ministers in May 1996, contains costed action plans for 14 key habitats and 116 species of plants and animals (49 of these depend on freshwater habitats). A UK Targets Group was set the task of producing action plans for a further 24 key habitats and 286 species by the end of 1998. A number of regional and local biodiversity initiatives are also being undertaken. The action plans set conservation targets and timetables for action. Key wetlands identified include fens, reedbeds, coastal and floodplain grazing marsh, raised bogs, blanket bogs and wet woodlands. Examples of management objectives and hydrological objectives required to meet these are given in Box 9.4.

9.6 Hydrological management of wetlands

Water Level Management Plans (WLMP) were initiated by the Ministry of Agriculture, Fisheries and Food to provide a means by which the water level requirements for a range of activities in a wetland (including agriculture, flood defence and conservation) can be balanced and integrated. Wetland wildlife is dependent on an appropriate water management regime to maintain the different degrees of wetness which are required or tolerated by different plant and animal species. Ideally, plans should be prepared for all areas which have a conservation interest and where the control of water levels is important to the maintenance or rehabilitation of that interest. The initial programme concentrates on Sites of Special Scientific Interest (SSSIs). However, no specific funding has been provided by the Ministry of Agriculture, Food and Fisheries (MAFF) and implementation relies largely on the existence of other incentive schemes, such as the Wildlife Enhancement Scheme or Countryside Stewardship which pay farmers a subsidy for implementing environmentally friendly farming practices and maintaining high ditch water levels. Many plans produced to date have been quite qualitative in nature and need to quantify water levels, which may require specific scientific studies. The Environment Agency, the Jackson Environment of University College London and the Institute of Hydrology are undertaking joint hydrological research on the Pevensey Levels, a wet grassland, to define an accurate water balance that will underpin the Water Level Management Plan being developed for these wetlands. Figure 9.6 shows the initial estimates of the water balance components. It can be seen that the water balance is dominated by rainfall and evaporation, with inflow to and outflow from the wetland being relatively small. Evaporation is particularly difficult to measure directly and indirect methods are not well defined for wetlands. On the Pevensey Levels, a Hydra is being used which measures evaporation by the eddy correlation method

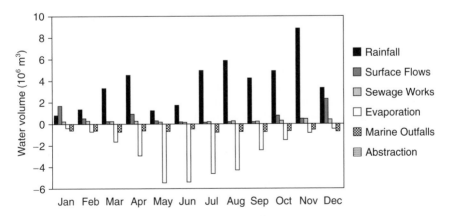

Figure 9.6 The water balance of the Pevensey Levels wetlands, Sussex, 1992.

(a)

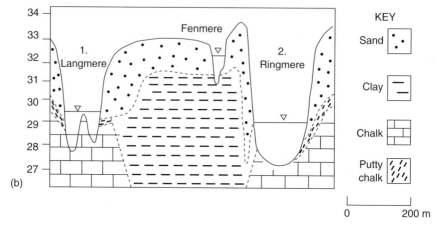

(b)

Figure 9.7 Breckland meres: (a) location; (b) geology.

(Shuttleworth *et al.*, 1988). It is hoped that this will allow development of correction factors to apply to standard meteorological measurements to estimate evaporation rates from wet grassland wetlands more accurately.

The interaction between wetlands and the other components of the hydrological regime within a catchment is often complex. In particular, the relationships between surface water, shallow groundwater and deep groundwater are complicated and difficult to quantify. In the case of wetlands, each is unique in some way, and it is difficult to extrapolate results of scientific studies from one site to another. For example, the Breckland Meres (Langmere, Ringmere and Fenmere) are three superficially similar endorheic base-rich wetlands on the Norfolk/Suffolk border (Figure 9.7(a)). Langmere is in hydrological continuity with the chalk aquifer (Figure 9.7(b)) and its regime is controlled by groundwater fluctuations. Ringmere has a less well developed connection with the chalk due to a lining of organic matter but is still largely controlled by groundwater. In contrast, Fenmere is separated from the underlying chalk by clay alluvium and its water level is largely a reflection of the balance between rainfall and evaporation (Denny, 1993). For some wetlands which are linked hydrologically to an underlying aquifer,

216

Box 9.5 Hydrological impacts of peat extraction at Thorne Moors

Thorne Moors in south Yorkshire is the largest remaining area of lowland raised bog in England. The whole mire has been subject to peat cutting since at least the seventeenth century and although the area has been designated as a SSSI, the landowner, Levington Horticulture, exercises pre-existing mineral extraction rights on the northern part of the moor, having donated the southern part to English Nature for conservation purposes. Conservation bodies were concerned that the hydrological integrity of the conservation area was threatened both by continuing peat extraction, and artificial pump drainage of the worked area, and also by regional groundwater abstraction for public water supply.

A hydrogeological survey and review by the Institute of Hydrology indicated that the site was underlain by thick deposits of boulder clay and pumping tests indicated that the permeability was so low that any downward leakage would be insignificant. Regional groundwater abstraction was therefore not perceived as a threat. Attention consequently centred on the potential for lateral drawdown of the moorland water levels towards the workings. A large-scale pumping experiment was designed to estimate the effective hydraulic conductivity of a block of peat some 100 m^3 in volume. This enabled both the matrix and the preferential flows to be taken into account, yielding a value one to two orders of magnitude higher than the mean of the previous measurements. This value was used in a groundwater simulation exercise to predict the likely watertable drawdown in the peat for different widths of buffer zone between the worked area and the conservation area, as a basis for discussion of the options for future site management.

---

their functioning may vary depending on prevailing conditions. For example, when the watertable is high the aquifer may be supplying water to the wetland (groundwater discharge), but as the groundwater level drops the hydraulic gradient reverses and the wetland may be supplying the aquifer (groundwater recharge). In this situation, the effects of groundwater abstraction will depend on the prevailing hydrological conditions.

Peat cutting and gravel removal are two practices which have impacts on wetland hydrology, changing water levels and pathways and the rate of drainage to or from surrounding land or water courses (Box 9.5). Nevertheless, extraction often creates lakes, fens, reedbeds and other habitats, although they may be scientifically of less interest than the original wetland. The Norfolk/Suffolk Broads, for example, are flooded sites of peat extraction dating back to medieval times, but now represent one of Britain's finest wetlands that is important for wildlife and recreation, (Broads Authority, 1997).

Much of the concern about wetland degradation has been focused on the impacts of reduced flows caused by water abstraction. However, problems may also arise due to inappropriate or lack of site management. For example, many wetlands are dynamic. In the natural situation, floodplains form a transitional sequence of habitats from continuously flowing backwaters and sidearm channels to oxbow lakes and temporary pools that are created, destroyed and recreated as a result of hydrological processes. Many wetlands naturally infill with sediment and vegetation to become gradually drier – colonised first by floating and emergent plants, then by fringing plants and finally by terrestrial plants. At Windsor Hill Marsh near Shepton Mallet, the Somerset Wildlife Trust has instigated scrub clearance in order to preserve the wetland flora of this SSSI, and grazing to keep down future scrub regeneration. Management of the wetland is consequently required to maintain a constant aquatic ecosystem. Current approaches to wetland management have had to focus on working with the constraints of existing drainage infrastructure and manipulated water levels rather than restoring dynamic natural processes. Such an approach may protect existing wetlands from further damage but treating the symptoms rather than the cause cannot be considered as a sustainable solution. Nevertheless, many managed systems have high conservation value; for example, water meadows, which are long established and accepted man-made systems requiring continual management.

9.7   Methods for hydrological management of wetlands

Box 9.6 describes a method developed by Williams *et al.* (1995) to simulate the effect of abstraction on an idealised wetland. However, recognising that so little is known about wetland hydrology, the method relates to a simple model of a wetland and therefore has few parameters. The hydrogeological parameters of interest are hydraulic conductivity (permeability) and specific yield of the underlying aquifer materials, including any semi-confining layers. The hydrological parameters include recharge estimations, potential/actual evaporation and flow characteristics. The lack of understanding of how these parameters may vary, and the ranges which may be ascribed to the parameter values has led to a general distrust of the use of numerical modelling to predict the influence of groundwater abstraction on surface water bodies (particularly wetlands). It almost seems that as more work is carried out less is 'known' and uncertainties dominate.

Newbold and Mountford (1997) combined the results of a range of studies to define the water level requirements of wetland plants and animals (Box 9.7). This relies on simple site observations to estimate the tolerance ranges of wetland plants in terms of depth of surface water. Soil water level requirements are all given for some plants but these are less reliable and not

---

Box 9.6 **Modelling the impact of groundwater abstractions on wetlands**

Williams *et al.* (1995) present a suite of methods which allow an assessment of the susceptibility of a wetland to a specific proposed abstraction, regardless of the amount of data available.

Emphasis is placed upon the construction of a simple but comprehensive water budget model, on a monthly timescale, that will predict the impact of changes in groundwater supply on a wetland. This spreadsheet model, named MIROS, can be used on sites with no local data but can also be refined to take account of such data that is or becomes available.

Prohibition of abstractions that would cause any fall whatsoever in water levels at wetland sites is not a realistic option. For a given wetland site, predicted changes in water level and flow, resulting from a proposed groundwater abstraction need to be evaluated by an objective procedure which must recognise the water needs of plant communities. A new well function was developed to quantify the effect of pumping on an idealised wetland. The analytical model characterises a wetland by its size, expressed as an effective radius, and a parameter which is a measure of the resistance to flow between the underlying aquifer and the wetland.

The Sum Exceedance Value (SEV) is used as a measure of the length and intensity of summer drought in wetlands soils. The SEV can be calculated from measured or modelled water level data on an annual or longer term average basis. Williams *et al.* (1995) use a combination of the MIROS model and SEV method to estimate maximum acceptable drawdown in the wetland.

Future development could incorporate the plant water requirement data recently published by English Nature (Newbold and Mountford, 1997) in place of the Sum Exceedance Value.

---

transferable. Furthermore, it only provides a single value, or range of values, of soil water level and depth and does not consider the tolerance of species to the duration of period under given thresholds. During droughts, water levels may fall below these critical values, but provided they do not persist and do not reoccur frequently, many species can recover rapidly. This work needs to be extended to include tolerance to durations under given thresholds.

Gowing *et al.* (1994, 1997) have taken a more transportable and rigorous approach to estimating the water regime requirements of wetland plants (Box 9.8) through modelling funded by MAFF. The work produced the SCHAFRIM model, though this not to date been taken up by practitioners.

Box 9.7 Water level requirements of wetland plants and animals (Newbold and Mountford, 1997)

This worked produced:

- upper and lower limits of tolerance to depths of surface water of 246 wetland plants
- hydrological and ecological requirements of ten wetland birds
- preferred water level requirements of:
    6 over-wintering waterfowl
    8 reedswamp birds
    6 species of dragon fly and amphibian

Box 9.8 Water regime requirements of native flora (Gowing *et al.*, 1994)

Gowing *et al.* (1994) have determined quantitatively the tolerance of many native species to drought and/or water-logging, by using hydrological modelling at twelve wet-grassland sites across England. These sites, studied over a ten year period, include groundwater wetlands, such as the Somerset Levels and the Broads together with flood meadows associated with rivers such as the Thames and its tributaries. The species' tolerances have been parameterised in such a way as to allow information to be transferred between sites with different soil textures. This work has culminated with production of a model called SCHAFRIM which allows the prediction of habitat suitability for wet grassland plant species, based on a hydrological model of watertable behaviour.

9.8 Functional analysis of wetlands

The interactions between water, soils and organisms in wetlands express themselves as important ecosystem functions, some of which can be beneficial to mankind. These functions include flood alleviation, groundwater recharge, nutrients recycling, storm protection and pollutant trapping (Table 9.1). Clearly not all wetlands perform all these functions; for example, reedbeds are particularly effective at removing nutrients, whilst floodplains are most effective at reducing flood peaks. For these key functions, there is well documented scientific evidence. However, the extent to which other wetlands perform other functions is less certain. Furthermore, some wetlands perform functions which may be considered to be unhelpful to mankind. For example, some headwater river margins are zones of runoff generation and can augment floods. A comprehensive review of the scientific evidence for hydrological functions (Acreman and Bullock, in preparation) is underway which will provide a benchmark on which to undertake future research.

Groundwater recharge and discharge are the most relevant functions to consider here. Many wetlands exist because of an impermeable layer, e.g. clay, which would exclude such processes. However, other wetlands are linked hydrologically to an underlying aquifer, allowing exchange between them. When the watertable is high, the aquifer may be supplying water to the wetland (groundwater discharge), but as the groundwater level drops the hydraulic gradient reverses and the wetland may be supplying the aquifer (groundwater recharge). The effects of groundwater abstraction will depend on the precise local hydrological conditions.

There is a need to develop generalising rules so that the magnitude of hydrological fluxes can be estimated from easily measured catchment characteristics and results are transferable between different wetland studies. Maltby *et al.* (1996) have developed a framework for the functional analysis of wetlands. Their work aims to build the possibilities of predicting wetland functioning by characterisation of distinctive ecosystem/landscape units (termed *hydrogeomorphic units*). Although the objective is a simple and rapid procedure for use by professionals who do not have an in-depth knowledge of wetland ecosystems and their functions, the system still needs to be developed into operational tools.

9.9 Wetland restoration

The growing realisation that wetland degradation has led to the loss of important economic and ecological values has led to a range of restoration projects. A good example is the work at Redgrave and Lopham Fen in Norfolk (Figure 9.8(a)). Compared with surveys undertaken in the 1950s, the vegetation communities were much degraded as a result of a drop in water level of more than a metre. The most well-known species on the site, the great raft spider (which only occurs at one other site in the UK, Pevensey Levels) was on the brink of extinction. In addition, many of the 125 Red Data Book invertebrates had shown significant decline. Hydrological and hydro-geological modelling studies showed that the main cause had been pumping of groundwater for public supply adjacent to wetland that began in 1961. This was confirmed by an experimental cessation of pumping in 1990 (Figure 9.8(b)). The borehole was subsequently relocated at a cost of £2.3 million. Desiccation of surface soil layers had allowed encroachment of terrestrial species. Trees were removed and turned to wood chips to fuel a local biogas plant and the top layer of oxidised peat is currently being stripped to expose saturation peat which will support wetland plant communities.

(a)

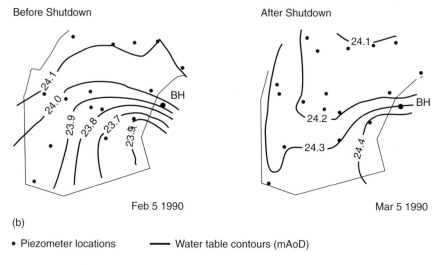

Before Shutdown

After Shutdown

BH

BH

Feb 5 1990

Mar 5 1990

(b)

• Piezometer locations   —— Water table contours (mAoD)

Figure 9.8 Redgrave and Lopham Fen: (a) location; (b) groundwater contours before and after experimental cessation of pumping in 1990.

9.10 Conclusions

Wetlands are important features of the British landscape, providing a range of important products (reeds, peat, hay) and values (walking, birdwatching, fishing, boating). Interactions between the water, soils and plants of the wetlands and rocks beneath them expressed themselves as hydrological functions, reducing floods, recharging groundwater and removing pollutants and sediment. However, to perform these important and valuable hydrological functions, wetlands must be carefully conserved. The UK has designated many UK wetlands as Ramsar sites, National Nature Reserves (NNR), Special Areas of Conservation (SAC), Special Protection Areas (SPA) and Sites of Special Scientific Interest (SSSI). For example, 27 UK wet grassland sites are

designated as SPA/Ramsar sites in recognition of their international importance for wintering birds (Benstead *et al.*, 1997). However, wetlands are still
being lost due to poor awareness of their important economic, cultural and
aesthetic value.

References

Acreman, M.C. and Adams, B. (1998) *Low Flows, Groundwater and Wetland Interaction.* Report to Environment Agency (W6–013), UKWIR (98/WR/09/1) and
NERC (BGS WD/98/11). Institute of Hydrology, Wallingford.

Acreman, M.C. and Bullock (in preparation) Hydrological functions of wetlands.
IUCN, Gland, Switzerland.

Barbier, E.B., Acreman, M.C. and Knowler, D. (1997) *Economic Valuation of Wetlands. A Guide for Policy Makers and Planners.* Ramsar Convention Bureau. Gland,
Switzerland.

Bateman, I.J., Langford, I.H., Graham, A. (1995) A survey of 'non-users' willing-
ness to pay to prevent saline flooding in the Norfolk Broads. *CSERGE Working
Paper GEC 95–11.* University of East Anglia, Norwich.

Benstead, P., Drake, M., José, P., Mountford, O., Newbold, C. and Treweek, J.
(1997) *The Wet Grassland Guide – Managing Floodplain and Coastal Wet Grasslands
for Wildlife.* Royal Society for the Protection of Birds, Sandy, UK.

Biodiversity Challenge (1996) *High and Dry: the Impacts of Over-abstraction of Water on
Wildlife.*

Broads Authority (1997) *Broads Plan 1997 – The Strategy and Management Plan for
the Norfolk and Suffolk Broads.* Broads Authority, Norwich.

Coles, B. (1995) *Wetland Management – a Survey for English Heritage.* Wetland Archaeology Research Project Occasional Paper 9. University of Exeter, Exeter.

Davis, T.J. (ed.) (1994) *The Ramsar Convention Manual: a Guide to the Convention on
Wetlands of International Importance Especially as Waterfowl Habitat.* Ramsar Convention Bureau, Gland, Switzerland.

Denny, P. (1993) Water management strategies for the conservation of wetlands.
J. IWEM. 7, 387–394.

Dugan, P. (ed.) (1990) *Wetland Conservation: A review of Current Issues.* IUCN, Gland,
Switzerland.

English Nature (1996) *Impact of Water Abstraction on Wetland SSSIs.* English Nature,
Peterborough.

English Nature (1997) *Wildlife and Fresh Water – an Agenda for Sustainable Management.* English Nature, Peterborough.

Environment Agency (1998) *A Price Worth Paying. The Environment Agency's Proposals
for the National Environment Programme for Water Companies 2000–2005.* Environment Agency, Bristol.

European Commission (1995) *Wise Use and Conservation of Wetlands.* COM (95) 189
final., Brussels.

Gowing, D.J.G., Gilbert, J.C., Youngs, E.G. and Spoor, G. (1997) *The Water Regime
Requirements of the Native Flora – with Particular reference to ESAs.* Report to Ministry of Agricultural, Food and Fisheries. Project BD0209. Silsoe College, Silsoe.

Gowing, D.J.G., Spoor, G., Mountford, J.O. and Youngs, E.G. (1994) *The Water Requirements of Lowland Wet Grassland Plants.* Report to Ministry of Agricultural, Food and Fisheries. Silsoe College, Silsoe.

Hanley, N. and Craig, S. (1991) Wilderness development decisions and the Krutilla – Fisher model: the case of Scotland's Flow Country. *Ecolog. Econ.,* 4, 145–64.

Haycock, N.E. and Burt, T.P. (1993) Role of floodplain sediments in reducing the nitrate concentration of sub-surface runoff: a case study in the Cotswolds, UK. *Hydrol. Proc.,* 7, 287–95.

Maltby, E., Hogan, D.V. and McInnes, R.J. (1996) *Functional Analysis of European Wetland Ecosystems. Phase I (FAEWE).* Final Report to European Commission EC DG XII STEP CT90-0084, Brussels.

Merritt, A. (1994) *Wetlands, Industry and Wildlife – a Manual of Principles and Practices.* The Wildfowl and Wetland Trust, Slimbridge.

Mitsch, W.J. and Gosselink, J.G. (1993) *Wetlands, 2nd edn.* Van Nostrand Reinhold, New York.

National Rivers Authority (1995) *The NRA's Role in Wetland Conservation.* R&D Report 351. National Rivers Authority, Bristol.

Neild, C. and Rice, T. (1996) *A Review of the UK Compliance with the Ramsar Convention on Wetlands of International Importance especially as Waterfowl Habitat.* Friends of the Earth, London.

Newbold, C. and Mountford, J.O. (1997) *Water Level Requirements of Wetland Plants and Animals.* English Nature Freshwater Series No. 5. English Nature, Peterborough.

Russel, M.A. and Maltby, E. (1995) The role of hydrologic regime on phosphorus dynamics in a seasonally waterlogged soil, in Hughes, J.M.R. and Heathwaite, A.L. (eds), *Hydrology and Hydrochemistry of British Wetlands.* Wiley, Chichester.

Shuttleworth, W.J., Gash, J.H.C., Lloyd, C.R., McNeil, D., Moore, C.J. and Wallace, J.S. (1988) An integrated micro-meteorological system for evaporation measurement. *Agric. Forest Meteorol.,* 143, 295–317.

Stoneman, R. and Brooks, S. (1997) *Conserving Bogs. The Management Handbook.* HMSO, Edinburgh.

Williams, A., Gilman, K. and Barker, J. (1995) *Methods for the Prediction of the Impact of Groundwater Abstraction on East Anglian Wetlands.* British Geological Survey Technical Report WD/95/SR, Wallingford.

Section 3

RESPONSES

10

RESPONSIBILITIES AND STRATEGIES OF UK ORGANISATIONS

Susan Walker, Bob Sargent and John Waterworth

This chapter considers the responsibilities and the strategies of UK organisations: the Environment Agency in England and Wales, the Scottish Environment Protection Agency in Scotland and the Department of Environment, Transport and the Regions in Northern Ireland. It also outlines the role of some of the other stakeholders in water management including the water companies, power generators, local authorities and research organisations.

10.1   Administrative arrangements

Within England and Wales responsibility for water management rests largely with the Environment Agency (EA), within Northern Ireland, the Department of the Environment, Transport and Regions, Northern Ireland Office (DETR NI) and within Scotland, the Scottish Environment Protection Agency (SEPA). The roles and remit of each of these three organisations is somewhat different, due largely to the underlying legislative and administrative differences. This means that there are also other major players in water management in different parts of the United Kingdom who may be responsible for some of the powers or duties which may be vested in one of these three major organisations in other parts of the Union. For example DETR NI is responsible for provision of public water supplies. In Scotland this is the responsibility of Regional Water Authorities and in England and Wales it is the responsibility of the privatised water companies. Hydrology underpins the water management activities in all these organisations. Indeed, a recent membership survey by the British Hydrological Society showed that 6% of UK hydrologists work from government departments, 16% work for regulators such as the Environment Agency and SEPA and around 10% to work for water companies.

10.2   Environment Agency

The Environment Agency was set up under the Environment Act 1995, bringing together the activities of Her Majesty's Inspectorate of Pollution, the Waste Regulatory activities of 83 local authorities, the National Rivers Authority (NRA) and several small parts of the Department of the Environment (DoE). The Agency is a non-departmental public body sponsored by Secretaries of State for the Environment, Transport and Regions and Wales and the Minister for Agriculture, Fisheries and Food. It carries out both regulatory and operational activities over a wide range of environmental functions across England and Wales. By bringing together the often fragmented activities of its predecessor bodies, the Agency is able to contribute to sustainable development, and to manage, protect and enhance the environment in a multi-functional and integrated way. This extends beyond the water environment by having an integrated approach to environmental management across air, land and water. The Agency takes a long-term view, basing its decisions on sound science and is prepared to take a precautionary approach where uncertainty exists and the consequences of getting it wrong are severe.

Overall, the Agency has a staff of over 9000 and an annual turnover of over £500 million. However, the Agency also works with others to deliver environmental improvements and to ensure that the existing quality of the environment is not degraded. This includes those who the Agency regulates, such as water companies and industry, local authorities who, for example, have a key role to play through the planning and development control legislation and environmental groups and the community at large.

Functionally, the activities of the Agency can be split into Environmental Protection and Water Management, though clearly there is some overlap between the two and a need to work together to deliver integrated river basin management. Environmental Protection encompasses regulating the disposal and carriage of waste, protecting and improving the quality of rivers, estuaries and coastal waters and regulating major industrial processes, nuclear sites and premises authorised to dispose of radioactive waste.

Water Management includes those activities previously carried out within the National Rivers Authority: Water Resources, Flood Defence, Fisheries, Conservation, Recreation and Navigation functions. The Agency, as set out in Table 10.1, is funded from central Government grants (known as grant aid), and income from its own charging schemes. Local government funds flood defence in the form of regional levies. The Agency charges for fishing and navigation licences. Pollution control charges are levied on those with effluent discharge consents, waste management licences and integrated pollution control and radioactive substances authorisations. Most of those who abstract water require a water resources licence from the Agency to do so. They are then required to pay an annual amount based on the annual volume

Table 10.1 Environment Agency income.

	Operational Receipts (£m)	Grant aid Contributions (£m)
Environmental protection	88.4	79.1
Water resources	86.7	
Flood defence	201.4	49.4
Fisheries	13.4	7.5
Recreation and conservation	0.5	7
Navigation	3.7	3.7
Total	394.1	146.7

Source: Environment Agency Annual Report and Accounts 1996–7.

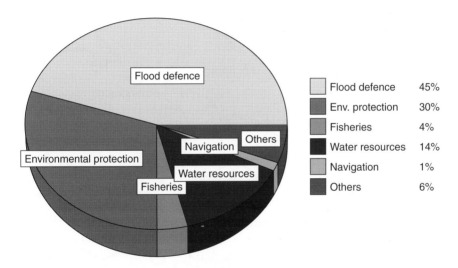

Flood defence	45%
Env. protection	30%
Fisheries	4%
Water resources	14%
Navigation	1%
Others	6%

Figure 10.1 Environment Agency expenditure
Source: Environment Agency Annual Report and Accounts, 1996–7.

licensed for abstraction. This pays for the Agency's expenditure on water resources. The income received in respect of any of Agency functions can generally only be used to fund Agency activities within that specific function.

Almost half the Agency's expenditure is spent on flood defence as indicated in Figure 10.1. Almost one-third is spent on environmental protection. The remainder is spent on Water Resources (14%), Fisheries (4%). The remaining functions account for 7% of expenditure.

The Agency is organised on a regional basis (Figure 10.2) with 26 areas in eight regions providing customer focused local service delivery. Local Environment Agency plans (LEAPs) are being developed in consultation

Figure 10.2 Environment Agency jurisdiction, boundaries and principal offices.

with industry, the community, local authorities and others. These identify local environmental issues and how they might be addressed whether by the Agency alone, or by the Agency working in partnership with others, or by other organisations such as local authorities exercising their statutory powers.

An understanding of hydrology underpins much of the Agency's business particularly in regard of Flood Defence, Water Resources and wider water environment issues.

Within Flood Defence, some £250 million annually are spent protecting people and property from flooding, by maintaining and improving river and sea defences and associated structures. Over three million people are protected together with property to the value of above £10 000 million. Some 1.2 million hectares of land are protected by 56 000 km of fluvial and 4400 km of sea and tidal defences. In addition to flood protection work, the Agency provides and disseminates flood warnings to the public by continually monitoring rainfall, tidal and river levels in order to predict when and where flooding might occur. When flooding is predicted to occur, an emergency workforce is available to provide emergency response and hence minimise the risk of flood damage.

Much of the flooding risk in urban areas is associated with increased and more speedy runoff associated with development. The Agency therefore works closely with local planning authorities to try to ensure new flood risks are not created nor existing risks made worse by inappropriate development activity. The Agency requires anyone wishing to carry out works which may affect a watercourse or flood defence to obtain a consent in order that it can satisfy itself that it will not constitute to an increased risk of flooding.

The main contribution of hydrology to the environmental protection function of the Agency is within the context of water quality with the Agency responsible for the water quality regulation over 42 000 km of river and canal, 2600 km^2 of estuary and 5000 km of coastline. Across England and Wales as a whole, there are over 100 000 water quality discharge consents. The Agency is responsible for water quality planning and for response to water pollution incidents.

The hydrological input to the protection of the air and land media is limited. However, hydrogeologists have a key role to play regarding the licensing of waste disposal sites and the remediation of contaminated land.

A major area of activity within the Agency for hydrologists is within the water resources function. The Agency is under the legislation required to conserve, redistribute and augment water resources, and to ensure the proper use of water resources. It is the Agency's role to balance, through the abstraction licensing system, the reasonable needs of abstractors within those of the environment. The Agency has a long-term water resources planning function by which the Agency seeks to ensure that the best use is made of existing sources before the new resources are exploited. This includes providing proactive demand management, leakage control and other waste minimisation initiatives. The Agency is also looking to remedy problems created in the past by working together with abstractors to reduce the environmental impact of existing abstractions.

The Agency is also very much involved within the protection of groundwater from both a quality and quantity perspective. Underpinning its water resources work and supporting other Agency functions is a comprehensive network of hydrometric stations.

231

Plate 10.1 A river flow gauging station (the Biel Water at Belton House) operated by the Scottish Environment Protection Agency (Mike Acreman).

10.3 Scottish Environment Protection Agency

The Scottish Environment Protection Agency (SEPA) was also set up under the Environment Act 1995 and it has many parallels within the Environment Agency in England and Wales. It also has some fundamental differences driven largely by historic legislation which have resulted in, for example, there being only a limited system for licensing abstractions in Scotland and for responsibilities for flood defence resting within local authorities rather than SEPA. SEPA is a non-departmental public body accountable to the Scottish Parliament. It was formed by the merger of seven River Purification Boards, Her Majesty's Industrial Pollution Inspectorate, the waste regulation and local air pollution functions of district councils and the river purification, waste regulation and local air pollution functions of the island councils. Elements of the hazardous waste inspectorate of the Scottish Office were also included. SEPA aims to provide integrated sustainable environmental management across the elements of the environment for which it has responsibility.

SEPA has a staff of almost 700 and an annual turnover in excess of £30 million. Of its income, about half comes from grant aid, the remainder coming from cost recovery charging schemes relating to environmental licences such as discharge consents and waste licensing. Like the Environment Agency, it works in partnership with other key organisations to deliver environmental improvements and avoid environmental degradation which it cannot deliver alone.

A high proportion of the staff and resources within SEPA are directed at the Pollution Prevention and Control functions and the provision of the technical aspects such as laboratories and suitable advice which support that function. Almost 90% of staff are employed in the three operational regions (Figure 10.3). Because of geography and the need to provide a locally based operational service, a further 20 local offices provide a focus for customers.

In regard of hydrology, the responsibilities of SEPA include hydrometric data gathering, provision of flood warning, some limited abstraction control and assessment of flood risk. In addition, of course, SEPA also has a duty to conserve and protect the environment using the principle of 'sound science' and basic information on water resources and the aquatic environment are clearly essential to undertake this work.

SEPA also undertakes and commissions research and development (R&D) projects in hydrology, many through the joint Scottish and Northern Ireland Fund for Environmental Research (SNIFFER). Current (1997) research projects of particular note include:

- long-term hydrometric variations in Scotland;
- prediction of acceptable river flows;
- development of low-flow estimation facilities;
- surface water yield assessment;
- development of a groundwater database.

SEPA's remit in regard of flood defence extends to the provision of flood warning and, a new duty under the Environment Act 1995, provides assessments of the risks of flooding. This latter responsibility requires close cooperation with local authorities to ensure a proactive approach to development control. SEPA has been especially active in the promotion of source control methodologies to ameliorate the impact of runoff from urban areas in relation to both potential flooding and environmental quality problems.

SEPA is responsible for protecting and enhancing the quality of 50 000 km of rivers and lochs, 800 km^2 of estuaries and over 7000 km of coastal waters. There are around 40 000 water quality discharge consents across the county. SEPA faces a major challenge in both effectively regulating existing consents and planning for the significant improvement in water quality needed to meet European standards.

As in England and Wales, the hydrological input to the protection of air and land is limited. Because of the nature of the geology, there is less need in many parts of the country for significant hydrological input to the regulation and licensing of waste disposal sites and remediation of contaminated land.

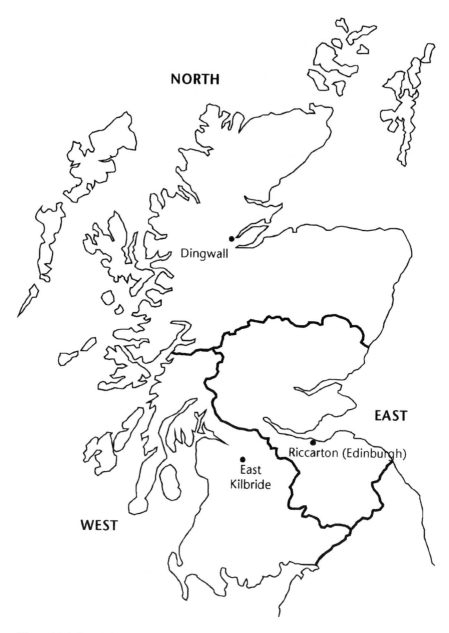

Figure 10.3 Scottish Environment Protection Agency jurisdiction boundaries and principal offices.

The contribution which can be made by hydrologists within SEPA to water resources management in Scotland is rather more limited than the case for colleagues in England and Wales because of the nature of the legislation north of the Border. There are only limited opportunities for licensing abstractions in relation to irrigation use and groundwater. Nonetheless, SEPA is charged with conserving water resources as far as is practicable, so issues such as yield of water resources, estimates of environmentally acceptable flows do have a relevance.

10.4   Northern Ireland

Since the mid-1970s, the two key departments in relation to hydrology were the Department of the Environment's DOE (NI), now DETR (NI), and the Department of Agriculture for Northern Ireland (DOE and DANI).

During the early part of 1990s, the two groups involved in surface water hydrology within DOE (NI) and DANI, began to examine the potential for co-operative arrangements leading to the elimination of areas of work duplication. A major force behind this was the focus given to the efficiency and provision of 'value for money' services by Government agencies. By 1996 both groups involved in hydrometry had gone through Prior Option Studies and emerged as Next Steps Agencies. The newly formed Environment and Heritage Service and the Rivers Agency are Departmental Agencies under the Next Steps Initiative.

Next Steps Agencies (NSAs) exist within their parent Departments but are no longer a part of the 'core' of the Departments. This distinction entails that a split between function and policy is in existence. In effect, the NSA is considered as a discrete business unit with responsibility for managing it's own functions and budget and headed by a Chief Executive. The creation of another NSA, the Water Service, with responsibility for water supply and waterborne waste treatment and disposal, has furthered the administrative transformation for Northern Ireland's Departmental bodies with an interest in and responsibility for hydrological monitoring.

The Rivers Agency responsibilities include promoting the conservation of water resources, cleanliness of waterways and underground strata, consenting of discharges and the collection/publication of data. DANI has discretionary powers among others to maintain designated watercourses and sea defences, construct and maintain drainage and flood defence structures and enforce the drainage function of all watercourses.

The transformation of organisational arrangements is not yet complete and the situation is still in a state of flux with a promise, yet to be fulfilled, that the need for both the Rivers Agency and the Environment and Heritage Service would be re-examined before the mid-term of the former. In effect, this could result in the merger of the two into a larger Agency closer in breadth of function to the Environment Agency of England and Wales. The

Water Service is facing a period of change and awaits the full results of a study of it's need for hydro-climatological monitoring and information in relation to water resource planning in the next millennium.

10.5 Water service companies in England and Wales

The 10 privatised water service companies were set up in 1989 when the existing Regional Water Authorities were split into the National Rivers Authority, as regulators and the privatised water companies as providers of water supply and sewerage services.

There is some overlap in the roles and responsibilities of the water companies and the Environment Agency, particularly regarding water resources and water quality management, where the technical issues of the regulator and the regulated are closely linked though their priorities and perspectives may be different.

Overall, water companies have only a handful of staff working as what might be termed hydrologists, though rather more work on water quality improvement schemes. Major hydrological drivers in recent years have been the 1995–96 drought and the need to recalculate the water resources/demand balance into the next century. A major interface between the water companies and the Agency is in the context of environmental planning and forward investment programmes for water resources and sewage and sewerage improvement schemes, and in the regulation of existing site authorisations.

10.6 Scottish water authorities

The three Scottish water authorities, North, West and East, were formed from the water and sewerage departments of the Regional Councils in 1996. They are non-departmental government public bodies and are responsible for water supply and provision of sewerage services in their areas. They provide hydrological information on rainfall and compensation flow. They also have a major role to play in assessing water resources and reservoir yields.

10.7 Local authorities

In Scotland, there are 29 unitary local authorities and three Islands councils who are responsible for local service provision. Some of these services have a hydrological element. The local authorities may construct and maintain flood prevention schemes in non-agricultural areas, and they can also undertake coastal protection work. Both of these provisions are subject to approval being given by the Secretary of State. Local authorities in England and Wales have less of an interest in hydrology than their counterparts in Scotland. However, a large number of them provide sewerage services to the privatised water companies on an agency basis. Local authorities are also responsible for coastal protection where there is some overlap and opportunities for collaboration within the Environment Agency's sea defence activities.

10.8   Power industry

The power companies, particularly Scottish Hydroelectric, provide hydrometric information for catchments used to generate hydro-electricity, and their operations can have a significant impact on the hydrology of some Scottish catchments. This is less of an issue in England and Wales. A significant amount of hydrological and hydro-geological work is being carried out by the nuclear power industry in support of investigations regarding the storage and treatment of nuclear waste.

10.9   Universities

Several universities have departments who are very active in hydrology. This work includes provision of undergraduate and postgraduate teaching in hydrology, hydrological research, consultancy work and the operation small research catchments. Some universities have significant databases of hydrological and hydro-geological data.

10.10   Meteorological Office

The Meteorological Office maintains the rainfall archive and collates other climatic data, using information provided by the Environment Agency, SEPA and others. They also provide forecasts of weather and tides which are used operationally by hydrologists, e.g. for flood warning and for water supply operations. The Meteorological Office carries out research into all aspects of meteorology including hydro-meteorology and weather radar.

10.11   Centre for Ecology and Hydrology

The Centre for Ecology and Hydrology includes the Institutes of Hydrology (IH), Freshwater Ecology (IFE), Terrestrial Ecology (ITE) and Virology and Environmental Microbiology (IVEM). It has its headquarters at Wallingford. The Institute of Hydrology carries out a wide range of hydrological research and consultancy into all aspects of the hydrological cycle, including studies of process and operational hydrology, hydrometry, hydroecology and climate change. It operates experimental catchments across the UK including Plynlimon (in upland Wales), Balquidder (in the Scottish Trossach uplands) and the Cairngorms mountains (Scotland). Here the work has particularly focused on the effects of land use, such as forestry, and on snow. In recent years, emphasis has moved to lowland impermeable catchments where water resources are particularly under pressure. The Institute of Hydrology also maintains The National River Flow Archive and the National Hydrological Monitoring Programme which provides status reports on the hydrological situation in relation to flood and droughts.

10.12   British Geological Survey

The Hydrogeology Group of the British Geological Survey (BGS) holds a significant amount of information on groundwater in the UK and has published widely on groundwater quality and resources. These publications provide a major source of information on this subject area. It also manages the National Groundwater Archive, a unique database containing over 130 000 borehole and well records, which is used as a basin for the Hydrogeological Enquiry Service. As a sister organisation to the Centre for Ecology and Hydrology, it provides a focus for research in hydrogeology outside the academic sector.

10.13   Need for legislative change

Even though there have been significant changes to water management legislation in the last 40 years, there are still a surprising number of legislative problems which inhibit the effective management of the hydrological cycle.

In England and Wales, there is currently ongoing (1997) a review of abstraction licensing. There are concerns that the current system, set up largely as a result of the Water Resources Act 1963, is inflexible in meeting changing demands on the water environment. For example, it has allowed situations to develop where abstractors acting entirely legally lower flows or even dry up watercourses because abstraction licences were set at too generous a level because either the potential environmental impact of the abstraction was unknown or not considered important at the time the licence was set. The existing legislation means that the Environment Agency can unilaterally change abstraction conditions under a licence only with the provision of significant monetary compensation to the abstractor. Likely outcomes of the review include some review process for licences and consideration of the deregulation of licences which have an insignificant environmental impact.

In Scotland, despite the long list of organisations involved in water management, there are significant legislative gaps. There is no overall control of surface or groundwater abstraction in Scotland. Abstractions for public water supplies have to be approved by the Secretary of State but the only provision for licensing of general water abstractions is for agricultural irrigation in areas approved by the Secretary of State (in 1997, less than 100 km^2 had obtained such approval).

Flood protection is also not well catered for in the administrative framework. This is generally the responsibility of riparian landowners, with two minor exceptions. These are the provision for local authorities to designate flood prevention schemes in non-agricultural areas (subject to approval by the Secretary of State) and the provisions of the Agriculture (Scotland) Act 1948, which provided for collaborative drainage areas to be established by the Secretary of State. Although several of these schemes are still extant, these provisions are now little used. Individual landowners do instigate drainage schemes, of course, but these are necessarily local and can cause more trouble than they solve.

The absence of a 'flood protection authority' is an aspect of a more general absence of an organised and integrated water management authority in Scotland. Because of this, there has been little move towards catchment management, although a couple of examples of collaboration among the large number of bodies involved has simultaneously shown that such integrated management is possible, desirable and very difficult.

In Northern Ireland, in legislative terms, the main impetus has been focused on the need to provide more information in order to assist compliance with EC Directives such as the Urban Waste Water Treatment and the Freshwater Fish Directives. Perhaps the most significant developments in this area are the amended Water Act (Northern Ireland), to replace the existing Act, and the Ecological Quality Assessment Framework Directive which identifies the measurement of quantity of water as a key parameter in the assessment of aquatic ecological quality. The new Water Act will strengthen powers to licence and charge for water abstraction. The knock-on effect of forthcoming national and international legislation will be to highlight and extend the role of hydrological monitoring in the protection of the aquatic environment and in the sympathetic development and protection of water resources.

Other pressures for change in Northern Ireland include the continued debate over the potential for Water Service to carry out it's function in the private sector. In the current local and national political climates, it is difficult to envisage this taking place, but there have been increased calls for the environmental regulator, which will contain a hydrometric group, to be located outside central government as some form of non-departmental public body. This structural change would have many advantages, such as a separation of the groups involved in discharging waterborne waste and protecting the aquatic environment, and a greater perception of independence on the part of the regulator. Despite the large number of interested voices and opinions regarding the increasing role of hydrology, the fragmented nature of administrative and functional arrangements within Northern Ireland will continue until due consideration is again given to the value of a co-ordinate, centralised approach.

10.14 The future

Hydrology in the UK has recently been and continues to be in a state of flux. The recent organisational changes in regulators and water authorities has initiated changes which have yet to be fully expressed in terms of hydrological activity and technological change. They will undoubtedly affect the future development of water resources management. The main drivers for change are the increasing development of legislation and the high priority given to the environment and to water resources management in particular by the Labour Government which came into being on 1 May 1997. There are large potential opportunities for the place of hydrology in the running of the country. Uncertainty is compounded by changes which may occur in organisational arrangements for water management following on from the election of Scotland and Welsh Assemblies and moves towards regionalisation in England and Wales. Further uncertainty relates to the forthcoming European Water Framework Directive. As yet, its potential implications are unclear.

Much of the impetus for change derives from the increasing realisation of the need to conserve and sustainably manage natural resources, of which fresh water is the prime example. The desire for development to be sustainable requires knowledge of the size and carrying capacity of water resources, whether these be freshwater flow in a river or groundwater in an aquifer. This need has been behind the development of hydrology in the last 40 years. What is new, however, is a realisation of the holistic nature of natural resources and this will take hydrology into much closer relationships with other sciences.

This change in emphasis is seen on the development of ecological flows, for instance. Where previously a steady compensation flow was considered to be adequate, the need is now seen for flow to vary to fit into the requirements of wildlife and to allow physical channel processes to continue. The flow regime is thus being considered as part of the ecological value of the riparian corridor, emphasising further the value of hydrological knowledge in conservation and management of the environment. Similarly, the effect of land use change on the hydrology of a catchment, and the effect changes in hydrology will have an all other aspects of the catchment will require integration of hydrology and the hydrologist into environmental management as a whole.

Also new is a wider acceptance by Government and the public at large that persistence of hydrological conditions cannot be taken for granted. This leads to a need to develop hydrology and water resource management and its legislation and institutional arrangements which can be flexible to changing needs in the future.

Appendix 1: Other UK organisations with responsibilities for water

English Nature

- advising the UK Government on the designation and management of sites under the international conventions and EC Directives (Ramsar sites, SPAs and SACs);

- establishing and managing National Nature Reserves (173);
- selecting and notifying Sites of Special Scientific Interest (3859 covering 934 400 ha);
- securing sustainable management of SSSIs in partnership with land-owners, occupiers and relevant statutory bodies;
- biodiversity conservation, including a Species Recovery Programme for endangered plants and animals;
- promoting conservation of the English countryside and its wildlife through a Natural Areas approach.

Deparment of the Environment, Transport and the Regions (DETR)

- implementation of EC Directives on water, habitats and species;
- designation of internationally important wetlands;
- leadership of biodiversity and sustainability programmes;
- sponsorship of English Nature, Environment Agency and British Waterways;
- water quality objectives and water protection zones;
- code of practice on conservation, access and recreation in the water industry and enforcement of water companies' environmental duties.

Ministry of Agriculture, Fisheries and Food (MAFF)

- flood and coastal defence, fisheries and agricultural policy;
- reducing farm pollution and promoting safe use of pesticides (e.g. Code of Good Agricultural Practice);
- water Level Management Plans initiative;
- agri-environment schemes for the countryside (e.g. Countryside Steward-ship, ESAs, Habitat Scheme);
- provision of Code of Practice on Environment Procedures for Flood defence Operating Authorities;
- control of introduction of non-native aquatic species.

Office of Water Services (OFWAT)

- sets eligible expenditure for water companies.

Forestry Authority

- promotion and advice on establishing new woodlands in floodplains and river corridors;
- best practice advice on wet woodland management including specialised techniques and machinery required;
- regulation of tree-felling in wet woods.

Countryside Council for Wales

- conservation of nature, landscape and amenity in the Welsh countryside – similar duties to English Nature and Countryside Commission.

Scottish Natural Heritage

- conservation of nature, landscape and amenity in Scotland – similar duties to English Nature and Countryside Commission.

Department of Environment, Transport and the Regions, Northern Ireland

- conservation of nature, landscape and amenity in Northern Ireland – similar duties to English Nature and Countryside Commission.

Joint Nature Conservation Committee

- advice on nature conservation (UK and international) on behalf of English Nature, Countryside Council for Wales, Scottish Natural Heritage and Department of Environment, Transport and Regions Northern Ireland.

British Waterways

- management of majority of canal network (plus reservoirs and some river navigation).

Internal Drainage Boards

- management of water levels in 235 internal drainage districts;
- flood defence schemes.

Broads Authority

- Recreation, conservation and navigation in the Norfolk Broads.

Landowning bodies (National Trust, Forest Enterprise, sand and gravel companies, Country Landowners Association)

- management of land holdings, including important wetland areas;
- implement agri-environment schemes;
- influence government policies on water.

242

Industry (individual companies and their associations, e.g. CBI)

- users of water;
- polluters, either directly or through their products;
- landowners and developers;
- sponsors of conservation initiatives.

National Park Authorities

- planning and management of recreation and land use in National Parks.

Voluntary conservation organisations (e.g. RSPB, Wildlife Trusts, CPRE, WWF, WWT, FoE, Plantlife)

- protect species and restore riverine and wetland habitats;
- survey habitats and record aquatic species;
- own and manage nature reserves;
- advise on management techniques (e.g. Rivers and Wildlife Handbook (RSPB), creation of industrial Wetlands (WWT));
- lobby for changes in government policy and legislation.

Angling and fishing associations and their representative bodies

- major users and potential guardians of our fresh waters.

Inland Waterways Association

- recreational use of waterways.

Inland Waterways Amenity Advisory Council

- advise on the recreational use of canals.

Royal Commission on Environmental Pollution

- carry out studies and make policy recommendations to Government (e.g. Freshwater Quality, 1992).

Countryside Commission

- conservation of amenity and landscape, including river valley and floodplain restoration projects.

11

PLANNING AND MANAGING
FOR THE FUTURE

Malcolm Newson, John Gardiner and Simon Slater

11.1 Introduction

River basins ('catchments' in the UK) are ideal environmental management units; they are clearly bounded, physically systematic, hierarchical in scale and culturally meaningful. However, it is only in recent years that the 'catchment consciousness', encouraged by scientists as early as Leonardo da Vinci in 1502, with his map of the Arno basin (Newson, 1992, 1997), has begun to overhaul the subsequent modernism of artificially-engineered, symptom and supply-orientated schemes of water management. 'Catchment consciousness' (Box 11.1) has led to new forms of debate about river and groundwater management (Gardiner and Cole, 1992).

Some consider that current 'water stress' (Slater, 1998) can be solved merely by the overlay of an economic system on the ecosystem (involving people via the free market, see Winpenny, 1994); the mainstream debate is, however, about democratic planning and interactive management as two practical means of ensuring sustainability through subsidiarity (Newson, 1997). Indeed, many have voiced demands for water management to take a lead in sustainable environmental management via an ethical stance (Feldman, 1991; Postel, 1992) and a new hydrological professionalism (Falkenmark and Lunqvist, 1995).

Despite this idealism there are still too many examples of how the fragmented nature of the institutional structure fails the community through taking too parochial a view of benefits and costs, when an overall consideration of the hydrological cycle, the environment and the needs of future generations would paint quite a different picture of needs and capacities. The UK Round Table on Sustainable Development (1997) has described water as a 'bounded resource', going on to admit that, 'current management of freshwater is unsustainable in some instances'. The efficiencies and environmental sensitivities to be gained from holistic approaches to planning and design, enabled by the application of sustainability principles and methods, promises to reveal the vital synergy between sustainability and development (e.g. von Weizsäcker *et al.*, 1977).

Box 11.1 **Indicators of 'catchment consciousness' in science and society** (Slater, 1997)

Hydrology. Identification of hydrological and environmental capacities and impacts of land use and development on water by catchment research

Hydraulic engineering. Imaginative use of technology, such as metering, leakage and source control, and recycling to assist in water management

Economics. Comprehensive use of environmental economics to assess new water schemes and the pricing of water and its use to manage demand

Society. Rise in public awareness of water issues within their catchment and involvement in education campaigns and local water decision making. Development and land use being accommodated, prevented or promoted according to its impact on the water environment and the current hydrological capacities

New policy framework. Use of the catchment as a flexible and adaptive framework for a mixture of management tools such as defining areas of water stress and appropriate policy responses including development allocation and accommodation

---

11.2   What future?

Whilst the terms 'water stress' or even 'water crisis' (EEA/UNEP, 1997) may smack of tabloid journalism, there are some essential issues to be resolved in UK water management; the urgency is not created by journalists or by the issues themselves (except, perhaps drought) but by an increasing public perception of 'ownership' of water issues, an atmosphere created partly by the politicisation and commodification of water since privatisation in England and Wales. Twenty years ago the question 'What future?' would have been answered purely technically as engineers responded to the challenge created by increasing human demand and (apparently) increasing natural variability in supply. Where to locate new dams? How big can we build? What new techniques can be applied to sewage treatment? These typify the water management questions of the 1970s and 1980s. By contrast, elements of the new debate about future strategy and operations include:

- Can that demand be managed or the impacts be mitigated at source?
- How can holistic approaches (e.g. integrated catchment management) minimise the demand?
- What are the effects of development on the human 'footprint' impacting on ecosystems?
- What are the real environmental sensitivities? How resilient is the system?

- How will changing climate and sea-level constrain our options and open opportunities?

We cannot merely record the change of ethos in our management of rivers; indeed, we need to understand and develop it further for an era of *existing* 'water stress' or 'water crisis' in many parts of the world and for the threat of *rapid change* in water resources and hazards during and following climate change (Arnell, 1996). The UK plays an important leadership role in international water affairs, exporting billions of pounds worth of expertise to the developing world annually. To date, this has logically followed the demand in the developing world for such items as hydrometric systems, dam building, structural flood defences and conventional, large-scale irrigation technology.

It is important for our technical competence to reorientate towards 'catchment consciousness' to reap the rewards of a rapid change of paradigm in water-based development (Bailey, 1996). Although it should be noted that consultancy firms were writing management plans for catchments in developing countries long before the practice took root in the UK, it is tragic that 'unthinking westernism' has, for example, introduced the water closet into some semi-arid countries to become a symbol of social rank, however inherently unsustainable.

It has become clear that decisions made in all sectors of development have more-or-less severe implications for the wellbeing of the water environment. There are few land processes, agricultural, arboricultural, industrial, commercial or domestic, which do not affect the quality or quantity of the hydrological cycle locally, regionally or internationally.

Allan (1996) has also added an international dimension which needs urgent consideration by UK consultants, that of '*virtual water*'. Whilst water is generally too bulky to move around in world trade, the agro-industrial products of 'green water' (Falkenmark, 1997) can be traded if investment in national water projects and schemes proves too expensive or unsustainable. The huge potential of the virtual water concept, however, currently remains locked up by the less imaginative politics of trans-national capitalism and world trade agreements.

11.3 How much do we know? Scientific basis for integrated catchment management

Leonardo's map of the Arno catchment came some 300 years before scientific hydrology gathered its first experimental data on a catchment basis. The twentieth century expansion of the science has been dominated by measurement systems for strategic and operational decision support (Wilby, 1996) but it has also seen a distinctive research field of 'land-use hydrology' yield substantive proof of the necessity to consider the land phase of the hydrological cycle as a dynamic system – sensitively affected by human activities,

Table 11.1 Land use hydrology: a sample of impacts of river basin development recorded in the UK.

Development	Impact on runoff	Impact on water quality
Field drainage	Increased 'flashiness' of runoff from peats and some clays; elsewhere increased baseflow, reduced peaks[1,2]	Increased leaching of nutrients and pesticides[3]
Arterial drainage/flood protection	More rapid downstream transmission of floodwaves; less floodplain storage[4]	Increased sediment concentrations in rivers[5]
Wetland loss	Loss of flood storage areas[6]	Reduced purification/ buffering of inflows[7]
Urbanization	Increased flashiness of runoff[8]	Increased sediment loads, storm sewer overflows, road runoff[9]

References
1. Robinson, M. (1985) The hydrological effects of moorland gripping: a reappraisal of the Moor House research. *J. Environ. Man.*, 21, 205–11.
2. Robinson, M., Ryder, E.L. and Ward, R.C. (1985) Influence on streamflow of field drainage in a small agricultural catchment. *Agric. Water Man.*, 10, 145–56.
3. Heathwaite, A.L. (1991) Solute transfers from drained fen peat. *Water, Air Soil Pollution*, 55, 375–95.
4. Robinson, M. (1990) Impact of improved land drainage on river flows. Institute of Hydrology Report, No. 113, Wallingford, UK.
5. Sear, D.A., Darby, S.E., Thorne, C.R. and Brookes, A.B. (1994) Geomorphological approach to stream stabilization and restoration: case study of the Mimshall Brook, Hertfordshire, UK. *Regulated Rivers: Res. Man.*, 9, 205–23.
6. Wilcock, D.N. and Essery, D.I. (1991) Environmental impacts of channelization on the River Main, Co. Antrim, Northern Ireland. *J. Environ. Man.*, 32, 127–43.
7. Baker, C.J. and Maltby, E. (1995) Nitrate removal by marginal wetlands: factors affecting the provision of a suitable denitrification environment, in J. Hughes and L. Heathwaite (eds), *Hydrology and Hydrochemistry of British Wetlands*. Wiley, Chichester. pp. 291–313.
8. Woolhouse, C.H. (1989) Managing the effects of urbanisation in the Upper Lee, in *Proceeding 2nd National Hydrology Symposium*. British Hydrological Society, London. 2.9–17.
9. Turnbull, D.A. and Bevan, J.R. (1994) Integrated water quality monitoring: the case of the Ouseburn, Newcastle upon Tyne, in Kirby, C. and White, W.R. (eds), *Integrated River Basin Development*. Wiley, Chichester. pp. 325–38.

principally urbanisation, forestry and intensive farming. An example of the complex connectivity is the history of Venice, its catchments, lagoon and flood risk (Penning-Rowsell *et al.*, 1998).

In the UK the early lead taken by the Institute of Hydrology, supported by university researchers and other institutes, has gathered the body of evidence linking land issues to water issues summarised in Table 11.1.

It is not axiomatic that such evidence, however strong, should be rapidly applied to amend the practices so as to benefit the water environment. There are many problems of extrapolation, verification and acceptance (Newson,

1995, 1997; Loucks, 1997), even within the agencies responsible for management and regulation of the water environment (Gardiner, 1997a). During times of an anthropocentric approach to water management, characterised by apparent plenty, there has been little need to abandon a remote, technocentric philosophy based upon water as an engineered utility like any other. However, we are now (and will continue to be) in an era of 'water stress' characterised not only by local and episodic shortages in the UK but by considerable public attention to:

- quality of life, including the aesthetic aspects of rivers;
- the needs of non-human biota;
- the needs of future generations under changed circumstances, including further urbanisation and climatic change;
- cost-effectiveness of management;
- public accountability.

Unfortunately, 'land-use hydrology' has been preoccupied with its credibility as a physical science and has not moved on to answer questions of ecology, economics and their interaction. Merrett (1997) writes of 'hydro-economics' but, unfortunately, whilst this is a structure on which to place flesh, it cannot be given life until the life sciences have further clarified the 'true' environmental capacities of elements of the hydrological cycle. How much water does the river ecosystem need to avoid lasting damage? Despite decades of enquiry throughout the developed world, scientific analysis and policy development remain mutually weak in the area of predicting and prescribing minimum acceptable flows (Petts, 1996; Evans, 1997). As with so many other elements of the environmental science agenda at the millennium, about some aspects we know sufficient but not in an operational fashion, whilst about others we simply do not know enough or are too uncertain of our results.

11.4 Sustainability and the catchment ecosystem

At present, therefore, our best conceptual approach – and one which many are seeking to operationalise – is to lay out a catchment-scale ecosystem model (Marchand and Toornstra, 1986) and establish that 'damage' has occurred in the form of removal of the *spontaneous regulation functions* (Figure 11.1, Box 11.2) such as wetlands (Gardiner, 1994). We can list recommendations (Newson, 1997) for precautionary actions, we can promote rehabilitation of clearly damaged elements of the system (Brookes and Shields, 1996) and we can seek prevention of the causes (Gardiner, 1994), but we cannot yet securely deliver a risk-based assessment as a basis for public policy. Policy-makers are now accustomed to uncertainties in science and have begun to adopt 'least regrets' approaches, of which the catchment basis for planning

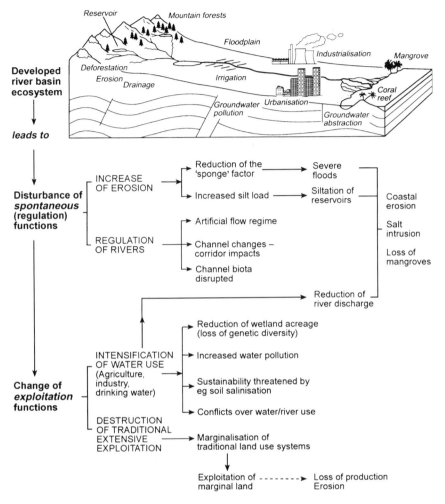

Figure 11.1 The catchment ecosystem (Newson, 1997; after Marchand and Toornstra, 1986).

and management is very prominent in many nations' policies (Box 11.3, Table 11.2). In the 'strongest' version of sustainable development (Pearce, 1993), protection of these spontaneous regulators would be seen as conserving critical 'natural capital'; a policy typical of this approach would be to encourage riparian buffer zones (Gardiner and Perala-Gardiner, 1997) within river corridor management (Gardiner, 1997b). Such approaches are complementary with the emerging holistic approach espoused by ecological economists (Savory, 1991). A group of UK public agencies has now promoted practical means by which environmental capital can be surveyed and valued (CAG/LUC, 1997).

Box 11.3 Advocates of the river basin as a planning and management unit

Quotation	*Reference*
• The integration of patterns and processes of both natural and social systems can be achieved in planning by using a watershed without neglecting or minimising the major elements of either system.	Hamilton and Bruijnzeel, 1997, p. 341
• Integrated water resource management, including the integration of land- and water-related aspects, should be carried out at the level of the catchment basin or sub-basin.	Agenda 21, Chapter 18 (see Johnson, 1993, p. 335)
• Planning will constitute a key mechanism to integrate all the components. The catchment or river basin should be the basic areal unit for such integrated man/land/water planning.	Falkenmark, 1997, p. 464
• It makes sense that the management of natural resources and the management of those human activities which are largely dependent on water should be undertaken within river basin units. . . .	Young, Dooge and Rodda, 1994, p. 19
• There is one point with respect to which both past policy and present policy are correct, and that is the paramount value of watersheds.	Leopold, 1924, reprinted by Flader and Callicott, 1991, p. 118
• . . . the idea of the drainage basin as an appropriate areal unit for the organisation of human activity and for regional planning has recently been revived with the recognition of the basin as an interrelated system.	Smith, 1969, p. 109

- From an environmental perspective, land and   Taylor, 1992, p. 97
  water would be the most important basic
  resources managed by the regions. These
  generate natural boundaries . . . referred to as
  'catchments' in geographic terms.
- The whole catchment area should be   UN Economic
  considered as the natural unit for integrated   Commission for
  ecosystems-based water management . . .   Europe, 1993, p. 1

Table 11.2 Selected international catchment planning experiences (Slater, 1998).

Country	Name	Earliest examples	Coverage	Scale	Land use links	Stakeholder involvement
Australia	Integrated Catchment Management	1970s	Partial	Local	Mainly rural	High
New Zealand	Catchment Plans	1940s	National	Regional	Mainly rural	Low
Canada	Watershed Plans	1970s	Partial	Regional and local	Urban and rural	High
USA	Watershed Plans	1930s	Partial	Regional and local	Urban and rural	Mixed
Portugal	River Basin Plans	1990s	National	Regional	Urban and rural	Low
Germany	Framework Plans	1960s	Partial	Local	Point source	Low
France	Catchment Management Plans	1920s	National	Regional and local	Urban and rural	High
Hungary	Regional Water Management Plans	1990s	National	Regional	Urban and rural	Low
Holland	Water Management Plan	1970s	National	Regional and local	Urban and rural	High
Italy	Water Catchment Plans	1970s	Partial	Regional	Mainly rural	Low
China	River Basin Development	1980s	Partial	Regional	Rural	Low
Nigeria	River Basin Plans	1960s	National	Regional	Rural	Low

The adoption of catchment planning is not merely an act of faith but follows the correct assumption that there are spatial and social elements to sustainable resource development, however sustainability is defined. Thus better resolution, an earlier 'answer' and a broader consensus on natural variability is produced at the relatively local scale, both strategically and operationally. Fortunately, UK catchments are an appropriately sized unit for this scale of approach; many of the world's water management problems need to be tackled across vast basins or aquifers, involving complex international security and culture issues (Gleick, 1993). In cases where the river catchments extend across distinctly different regions, natural resource management based on individual regions – and perhaps finding expression in sub-catchments – is practised (Hooper, 1997). Implementation of the European Union Water Framework Directive (European Commission, 1998) may be used to create a similar situation in the UK. Subsidiarity in decision-making not only advances democracy but incorporates the subtle spatial variation in risk-taking strategies which human societies derive from their indigenous knowledge and common cultural history.

The most frequently cited policy guidance to sustainable development since 1992 has been *Agenda 21*, the major strategic document stemming from the Earth Summit of that year; it has had a very successful local uptake all over the world. The National Rivers Authority Thames Region incorporated its principles in *Thames 21 – a Guide to the Resources and Development Pressures Affecting Water Management in the Thames Basin* (NRA, 1995). This consultative document uniquely identified the pressure points for the Thames basin water environment arising from future development proposed in the Regional Planning Guidance issued by the Secretary of State in March 1994. Its final form, issued in September 1995, was widely hailed as an important step forward to the inclusion of the several functions of the water environment in strategic planning, within one publication. The sequel, *Thames Environment 21*, launched in March 1998, takes a further step by applying a Strategic Environmental Assessment methodology to the pressure points identified in *Thames 21* (Environment Agency, 1998). One of the major goals in this process is identification of environmental carrying capacities and critical natural capital, using a credible methodology for their spatial representation on maps, to influence land use plans and policy statements as well as legislation and economic incentives (for land uses outside the planning system).

Despite the influence of the reinforcing messages of sustainable development (including a prominent role for water) contained in *Agenda 21*, it has proved politically difficult to force home a unifying definition and programme throughout the world. Yet this diversity, if not the result of political apathy, is to be expected; there will be 'versions' or 'strengths' of sustainability appropriate to each nation or even each catchment. The prominence accorded to ecosystem functions in any working definition of 'sustainable development' is said to define the 'strength' of that definition. The

Table 11.3 Sustainability in water management: 'strength' of sustainability (modified from Pearce, 1993).

	Policy	*Economy*	*Society*	*Contact*
Ultra weak sustainability	Lip service to integrated policies	Minor tinkering with economic instruments	Little awareness/ media coverage	Corporatist discussion, consultation exercises
Weak sustainability	Formal policy integration and deliverable targets	Substantial micro-economic restructuring	Wider public education – future visions	Round tables, stakeholder groups
Strong sustainability	Binding policy integration, strong international agreements	Full economic evaluation; green national and business accounts	Curriculum integration; local initiatives	Community involvement in decision-making

allocation of critical natural capital for protection is a characteristic of 'strong' sustainability but there exists a broader, evolutionary agenda for ecosystem entry into political decision-making, as outlined by Pearce (1993) and shown, for river management in the UK, in Table 11.3.

11.5 Why planning? How's it going?

Sustainability, however fully adopted, promotes a planning approach in the sense that planned occupancy of land seeks to apportion 'fair shares' of resources in an efficient way; carrying capacity arguments associated with strong sustainability operate in a similar way. Ideally, planning also seeks to anticipate future changes in the controlling variables, not just the ordered utilisation/conservation of the capacity. The UK has a considerable history of the involvement of the public in planning and this democratic tradition is ready-made for subsidiarity in water management; the public is, as we have said, pre-politicised for this addition to their democratic role and the link-ages between land use and water use made by catchment-scale planning is a considerable 'win' for land use hydrology and its results (summarised above in Table 11.1). In many ways catchments are in a leading position, inter-nationally, in moves towards operational use of environmental management units (even though most progress remains at the strategic level). It is signifi-cant that, in Australia, despite the well-developed regional resource man-agement and local community activity through Landcare groups, there is a generally perceived need for UK-style Catchment Management Plans and liaison with Local Planning Authorities. Australia has also adopted a 'bottom-up' approach to assessing ecosystem health at the catchment scale (Walker and Renter, 1996).

However, despite this rosy optimism, we must consider the daunting practicalities of institutionalising a successful scheme of catchment management planning in the UK. Immediately, we must consider the vital linkages between the agencies controlling water management, supply, re-cycling and those conducting strategic and development-control planning. We must also consider the bleak truth that, whilst we call the planning process 'town and country planning' it is highly restricted to urban land uses (Haslam and Newson, 1995); the rural dimension is controlled centrally (or in Europe) and by fiscal switches and agriculture policies.

Within the water field we must examine whether the seven main functional divisions of management (resources, flood defence, pollution, fisheries, conservation, recreation and navigation) can adopt a planning approach with equal facility and experience (Slater, 1997). On the face of it, in England and Wales, each function has its own traditions of adopting catchment outlines, relating to land use, employing planning liaison with local authorities, relating to the public and responding to environmental pressures. Of paramount importance is the level of scientific information available to each function and hence its *cultural* attitude to sustainable development. For example, in England and Wales, the *flood defence* function has long interfaced with land use for obvious reasons and is well supplied (albeit not spatially) with design criteria; it has its ecosystem protection options fairly well laid out. By contrast, *pollution control*, whilst aiming to use the spatial variability of ecosystem capacities (Newson, 1991) has tended to put the demands of human consumptive use uppermost. *Water resources* allocation has, to date, been similarly poorly equipped with the necessary science to respond to ecosystem needs.

There are also spatial variations which impact on the adoption of planning strategies – water resources in Scotland have never posed a threat to development opportunities whereas many would argue (NRA, 1994) that southeast England is now growing towards its supply limits. The big issue is, however, whether the seven management functions embodied in, for example, the Environment Agency can work jointly within an holistic framework towards the necessarily unified picture demanded of land-use planning (Ellis and Gardiner, 1998). The new Function Action Plans (Environment Agency, 1998) take a few steps towards this objective, but the traditional functional approaches still dictate their own agendas. There is no action plan for the integrative function of planning liaison or Local Environment Agency Plans (LEAPs) (reflecting the prevailing attitude that these are mainly administrative tasks), and little sense of real partnership with stakeholders to negotiate land use or the planning and management of its infrastructure for increased sustainability. As an acid test, no direct mention is made of source control, for example, an issue of deep interest to the Agency (Gardiner, 1998) and being promoted by forward-thinking Agency staff.

11.6 Catchment Management Planning: a 'LEAP' forward?

In England and Wales there is a significant problem of draining land and protecting it against flood damage (Figure 11.2); with the exception of rights to

Land areas dependent upon complete
systems for flood defence and land drainage

Figure 11.2 Areas of England and Wales dependent on flood protection and land drainage (from Newbold *et al.*, 1989).

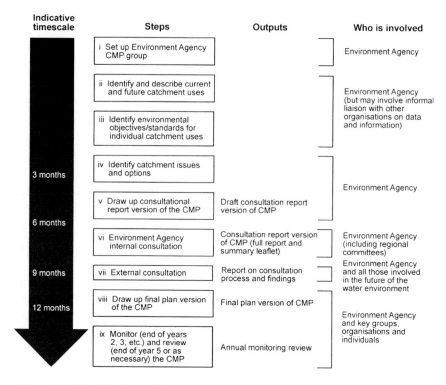

Figure 11.3 The Catchment Management Planning process as developed for England and Wales by the National Rivers Authority. (Modified from NRA, 1991.)

fishing, these aspects of river management have historically forged the major links between land and water. It is therefore no surprise that one aspect of 'water stress' emerging (predictably) in the Thames basin in the 1970s and 1980s became the basis of a new wave of 'catchment consciousness'. The Catchment Management Planning process is illustrated in Figure 11.3. It was a considerable coup for the NRA that, following the creation of the Environment Agency in England and Wales by the Environment Act 1995, catchment outlines broadly survived and grew in importance with the launch of LEAPs, in which strategies to solve problems of waste management and the majority of pollution control were incorporated.

By 1989, the National Rivers Authority began laying the outline basis of Catchment Planning and in its Corporate Plan for 1992–3 (NRA, 1991) launched the process with the following aims:

- integrating functional policies, providing information overview, consultation;
- prioritising needs, creating partnerships, linking land and water management;

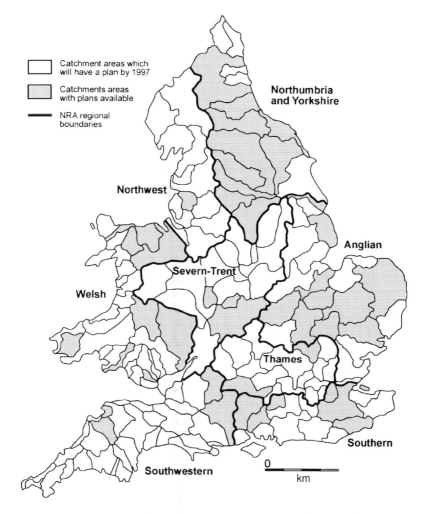

Figure 11.4 The division of England and Wales into catchment planning areas. (After NRA, 1995.)

- sustainable development, investment planning, remedying causes of problems;
- facilitating interdisciplinary action, guiding others' policies, consensus;
- forum for all users, tackling conflict, locating remedies.

Plans (later Catchment Management Plans) were to be drawn up for 164 catchments (Figure 11.4), with an average size of about 350 km^2. By 1996, when the NRA grew into the Environment Agency (EA), some 110 Plans had been completed.

The Plans and other documents in the process have been given considerable internal scrutiny within NRA and the Environment Agency, but the only comprehensive external review process has been conducted at the University of Newcastle upon Tyne (Slater *et al.*, 1995; Slater, 1997). At an early stage of the NRA CMP process, Slater *et al.* (1995) identified four major problems:

- insufficient spatial information in the consultation documents;
- lack of determined liaison with Local Authority planners (later partly rectified by the NRA itself (NRA, 1995);
- an unstructured approach to public consultation (a point picked up by the pressure group CPRE, Newson, Slater and Marvin, 1996);
- imprecise targeting (resource-based) of the catchment issues and no format for auditing progress.

Slater (1998) addresses two further scales of problem: the relationship between catchment issues and national policy development (including measures for regulating the private sector and national/regional planning guidance) and the lack of an holistic multifunctional approach. Space here allows us only to deal with the differences encountered between the three major water policy functions: water quality, water resources and flood defence (Table 11.4). These differences were brought out by intensive reviews of the planning process (as a whole) in three catchments in England: the Stour, the Kennett and the Bristol Avon (Slater, 1998).

Planned improvements in water quality can only operate in a fiscal context set by the financial regulation of the privatised water utilities (the same is partly true of water resources); local authority planners are also partly circumscribed by the attitude of central government to their pollution control responsibilities. Nevertheless, Slater's review of the Stour catchment suggests that catchment planning helps raise awareness of the spatial pattern of pollution, including the need for groundwater protection.

Put very coarsely, planning of water resources can only occur once the availability of surplus (low-risk) supplies becomes a location factor for development; regional patterns of population pressures and water availability are highly variable but there appears a political unwillingness to 'move people to water' when our technological history involves moving water to people. The planning picture, as shown by Slater's sample of the Kennet catchment, is further complicated by the inchoate approaches being used to control demand (and leakage from supply networks!) and the insufficient information available on environmental demands for water (and the benefits of supporting them).

From fieldwork associated with the CIRIA Project 555: Sustainable Urban Runoff Management, it appears that decisions on the drainage of major new developments are being taken without regard for the overall hydrological cycle, the economic or resource costs of extra abstraction, or the expansion of

Table 11.4 Parallel but distinctive paths: three main functions of river management in England and Wales – relationship with development planning (from material by Slater, 1998).

Issue	Water quality	Water resources	Flood defence
Pressures – emergence of 'water stress'	Urbanisation, agricultural intensification, low flows, new EU standards. Environmental capacities hard to define.	Recurrent drought, urbanisation/development in water-scarce areas. Hydrological limits known in 'average' conditions; ecosystem needs not well understood.	Impact of surface runoff from urban sources; demand for urban flood defences; rising sea-level; increasing storminess. Hydrological design and cost-benefit well established.
'New' hydrological information	Land use hydrology now legitimised but areal specification remains difficult e.g. for protection zones in diffuse sources.	Upland forest/urban impacts well researched but others (lowland forests, agriculture) neglected by catchment scale research.	Knowledge of e.g. hydrograph impacts well established since 1970s. Floodplain flow less well understood. Spatial variability of n-year flood still contentious.
Failures of traditional responses	Continued expansion of e.g. STWs politically unacceptable; increasing environmental costs; threat to environmental capacity e.g. eutrophication.	Few suitable reservoir sites and long lead-time for development approval. Transfers energy-inefficient and environmentally unacceptable. Public demands for reliability.	Large-scale drainage or flood defence schemes often politically resisted. Costs mounting; risk assessment difficult in changing climate/sea-level.
Development response	Development controls mostly at point-source or river corridor scale. Phasing of new development with STW upgrades in some areas.	Resource and development precautions still occur at national or regional level, not catchment scale for Plans. Phasing of development with supply and imposition of DSM techniques occurring in few places.	Responses remain largely at river corridor level, with extension to n-year floodplain where mapped. Some source area zoning and drainage plans. Strong line with new developments, e.g. zero runoff impact for developers.

Table 11.4 (cont'd)

Issue	Water quality	Water resources	Flood defence
Governance	Legislation enforced only since late 1970s; point sources; protection of private interests but also 'polluter pays' principle. DoE guidance on planning role in pollution control has been unclear.	Legislation followed national planning in 1960s; financed by loans and now WPLC income. Technical fix by engineering schemes has ruled; WPLCs have duty to supply and DoE guidance on development in relation to supply has been unclear.	Legislation from 1930s based on catchment outlines, yet finance for drainage was unlimited and not linked to downstream flooding. Reliance on 'hard' structural schemes until recently. Extensive and clear liaison with local planners and very clear DoE guidance since 1990s.
Demand side management and mitigation of development impacts	Protection zones and source control techniques now established means. Natural regulators such as wetlands increasingly valued. Incentive charging now proposed (c.f. cost-recovery).	Abstraction charges and metering still debated; 'smart' techniques under development. Consumer education popular following 1976 drought.	Source control techniques emerging, together with limited protection of natural regulators such as wetlands and floodplains.
'New' framework provided by CMPs and LEAPs	Internal stakeholders Better liaison has lead to priority action/response to AMP and to development plans External stakeholders Local authorities have improved information on e.g. zoning, source protection, diffuse sources. WPLCs have better response to AMP process Consensus building CMP failed to incorporate e.g. SWQOs.	Internal stakeholders Supply-led responses slowly being modified; much uncertainty about future climate; CMPs weak on water supply issues. External stakeholders Local planners found lack of clarity on resource issues; WPLCs found catchment scale information useful for investment planning. Consensus building CMPs helped start a debate about local supply issues but too often controls were national.	Internal stakeholders Flood defence originated CMP and strong planning liaison (varies regionally); highlighted the urban development issue. External stakeholders Local planners require updated floodplain maps and Standards of Service information. Public and pressure groups broadly satisfied – new partnerships with conservation. Consensus building Provides a framework for spatial debate about flooding.

centralised sewage treatment required, and certainly without any thought of the alternative of on-site treatment and recycling. The proximity of a pipeline seems to preclude any application of sustainability principles, or achievement of economic benefits which would be revealed by a more holistic approach.

Not surprisingly, therefore, it falls to the flood defence function to lead the way in two senses: that of influencing the surface water management of development which might exacerbate flood runoff from source areas, and also in controlling the location of development in flood-prone areas. In the latter case, the use of planning liaison is well-developed and supported by Circular 30 of 1992 and the subsequent Memorandum of Understanding between the NRA and local authorities. In the former case, the utilisation of planning law, initiated by Thames region's production of 'model policies' in 1989 which were adopted as 'model guidance' by the NRA, has so far avoided the need for separate legislation on controlling runoff. A kind of spatial demand management and mitigation of harmful impacts through negotiation has been in place for some time, as exemplified by Slater's study of the Bristol Avon, although it is better developed in some regions than others.

With the incorporation of much broader pollution control and waste management duties in the LEAPs, there exists a further potential for the legacy of functional attitudes to over-write the language of holism and for the traditions of the three agencies brought together in the Environment Agency to be divisive (in operations if not in strategy). Already, the use of catchment boundaries has been brought into question by, for example, waste management strategies.

Perhaps the biggest gap in both CMPs and LEAPs is in the area of *vision for changing conditions*. Slater *et al.* (1995) suggested that only one Plan which they consulted had a coherent vision for the catchment within a national policy context and in response to external (climate) changes. This partly reflects the lack of regional detail in developing scenarios for the impacts of change, but clearly the issues of water resources in southern England, flooding further north and changed water use by crops are now significant enough to be treated as issues at the strategic level. Possibly this particular gap is also indicative of yet another, namely that of calculating local environmental capacities and the sensitivity/resilience of catchment ecosystems. The 'ecological footprint' approach to water management would be a much more challenging approach to managing future change than is a list of 'catchment issues', however far-sighted and however frequently revised. It is significant that the Environment Agency has now launched a research and development initiative under the heading of 'catchment capacities'. Seeking to understand a catchment's dynamics, with a view to taking precautionary, preventative action, will minimise the footprint both ecologically and economically; some plans are beginning to embrace the 'generic issues' identified by Gardiner (1996) to this end.

11.7 Operationalising sustainability: the management challenge in changing conditions

Clearly, there are problems in putting all our institutional weight into planning scenarios. Although planning appears to be the most flexible strategy to cope with environmental, economic and social change, it faces a considerable challenge in the form of its knowledge base. Slater (1997) identifies the need for specification of sensitive zones on maps before management options can be considered. Those options seldom remain purely within the planning arena – technological fix is *not* a dead duck, especially in a nation suffering the consequences of a deteriorating Victorian water infrastructure. Monitoring needs to be constantly uprated as technology improves – real-time approaches to environmental capacities should be a possibility (e.g. pollution control) in the next millennium. Other technological developments will enable more reuse and recycling to diminish the effect of increasing demand. Change must be detected efficiently for precautionary action to be updated. Regulation and the use of economic instruments (especially for demand management) have a bright future, especially if they too can adopt a spatially varying, temporarily flexible framework. Figure 11.5 attempts to show where the institutional opportunities lie; in this scenario, the split private/public nature of water management appears to have considerable advantages in weaving a fabric of mixed regulatory, economic and planning approaches. Recent surveys have revealed the extent to which consumers are willing to accept regulatory costs for environmental improvements via their water services bills.

Our theme in this chapter is change; it is essential, if planning is to be a theme and a thread for (albeit 'weak') sustainable development, that it adapts to change. Planning is a process, not a document. The legendary military

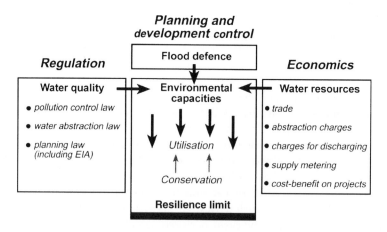

Figure 11.5 Regulation, economic instruments and planning: pathways to the utilisation and protection of environmental capacities.

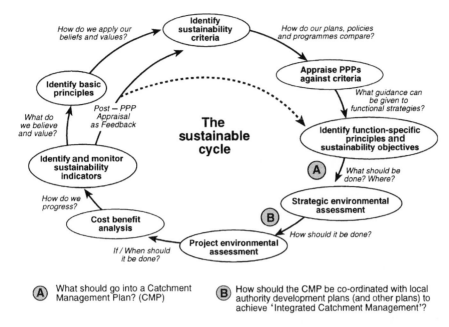

Figure 11.6 Sustainability cycles. (after NRA, Thames region, 1995)

failures of history often demonstrated the fallibility of plans made with poor information, yet rigidly adhered to! Immediately, therefore, there is a major role for science: – to inform the planning process with sound (not necessarily exact or certain) guidance. If the precautionary principle is to mean anything in operational terms, 'scenario setting' is a major responsibility for the hydrological research community. Paradoxically, the planning ethos is open to all science, including 'vernacular' science, or the 'common knowledge of ordinary folk'. Adaptive planning, like public response to hazards, requires options and experience; whilst the scientist may set up the valid options, it remains the experience of (and 'comfort' with) those options by the public which allows on-line adaptation of the plan to occur.

Now that Strategic Environmental Assessment (SEA) is gaining political acceptance and seems likely, together with river basin management plans, to become a statutory requirement, we may be on the verge of realising the iterative nature of the whole, stakeholder-driven process. This can be characterised by the seven-stage, sustainability cycle illustrated in Figure 11.6. It involves re-examining beliefs, values and hence basic principles, in the light of monitoring past projects against sustainability criteria, and then reviewing functional policies, plans and programmes against sustainability criteria to derive function-specific principles. These can be reworded as functional sustainability objectives, which can then be used to inform land use plans and enable SEA between functions and sectors to decide what should go where. How it should be implemented is then the subject of project environmental assessment, and the cycle begins again (Gardiner, 1996).

Whilst such structured, participatory planning may become our major device for coping with future uncertainties, the past is all too certain and obvious in many parts of the developed world. It is a past of unsustainable actions, leaving us with 'clean ups' and restoration as vital ingredients of environmental management. In the USA, home of the large dam philosophy of river management and the Tennessee Valley Authority, a programme of dam demolition (Shuman, 1995) has begun to gather pace. In Europe, the activities of river restorers has gained official backing and recognition. The River Restoration Project has successfully changed the landscape of 4 km of previously engineered channels in two UK sites, the Cole in Wiltshire and the Skerne in Co. Durham. Of note in both cases is the huge expense incurred by design and monitoring schemes connected to the restoration activity. However, these have been demonstration projects and such costs can be saved in future. Nevertheless, the two sites stand as tokens of an official intent to restore rivers and as a sign that this cannot be a strategy for all rivers.

Clearly three major forms of discipline are required. Firstly, to prevent damage to river ecosystems in the first place, i.e. to have an active conservation movement, backed by legislation and finance. Secondly, we need means of prioritising action, saving the most costly remedies for those situations where the need or likely outcome justify the means – River Habitat Surveys (Raven et al., 1996) are essential for this purpose. Thirdly, and more provocatively, it behoves those promoting sustainable 'back actions' to set more clearly the objectives of 'restoration'. On the ground, there are clearly at least three options which have been followed: enhancement, rehabilitation and restoration. These are clarified by Brookes and Shields (1996); Sear (1994) makes it clear that restoration sensu stricto is a catchment-wide agenda if the geomorphological basis for river dynamics is accepted.

11.8 Conclusions

The authors believe that planning and managing our resources of land and water within catchment outlines will form a major plank in sustainable development. However, others (albeit a minority) do not share this view. For example James Winpenny (1994) includes the following quotation in his work on the economics of water:

'The point is that geography and hydrology do not necessarily define the best scale for planning and problem-solving. Nor do they justify the use of 'integrated' or 'comprehensive' plans for water development'.

What we have tried to show is that regulation *and* economic instruments, used in conjunction with catchment planning and management, which will form a template for water management in future, on an international scale. An active awareness of the need to conserve the functioning of the hydrological cycle, and to protect the environment served by it, is a vital asset for those involved with decision-making over land use issues which have an impact on it. This means that the various effects of the proposal on the hydrological cycle should be identified spatially and in terms of quantitative and qualitative impact, together with possible mitigation of the effects, and alternative options. Similarly, decisions will be made in the context of a new dimension: the option of managing human demands for water (Newson *et al.*, 1996). In the cauldron of decision-making which characterises recent water management in South Africa an international statement, the 'Hermanus Declaration' (Environment Agency, 1997) is of special note. It includes the indictment of past models for water management as follows: 'There is considerable evidence from the international experience that the supply side option has subsidised over-consumption'.

Demand management, as a water-planning strategy, makes considerable sense in the UK given some of the scenarios of change already being outlined: more than two million new homes in drought-affected areas of England, a doubling of irrigation demand in East Anglia and a resulting doubling of the number of conservation sites under threat from over-abstraction (House of Commons Environment Committee, 1996; Stansfield, 1997). One aspect of demand management (research on drought-resistant crops) is already underway (Leeds-Harrison and Rounsevell, 1993) but the rise in the ratio of peak demand to average demand (to around 1.5) has not yet been reconciled with the water suppliers' claim to 'standards of service' to consumers which cannot be infringed – the environment suffers first, through drought orders.

For too long water has been ignored or made a subservient factor to economic and social considerations in development; the competence of civil engineering has been the 'tide to float all ships' in responding to human demands. Climate change, hydrological limits and the cost of traditional technical solutions are beginning to demand that water is an important *environmental* consideration of equal weight to social and economic drivers. Indeed, those projects or operations not based on the new sustainable development criteria of the 'triple-bottom-line' (including social and environmental considerations with economic decision making) face failure with dramatic consequences in the long term.

The problem is that elevating water to its rightful place in the economic and social decision making process faces a series of challenges:

- the dominant water management culture is regulatory and engineering based;

- the knowledge-base for evaluating the impact of water use and interaction with the environment, social and economic issues, although growing, is still patchy;
- the translation of risk concepts to the sustainable water environment as spatial data and economic analysis is still in its infancy;
- many of the sustainable solutions require action by the individual who has tended to be alienated by the technocratic control and now privatisation of water.

Catchment Management Planning and now LEAPs offer an opportunity to overcome these challenges by:

- providing a framework within which to carry out further research and test a variety of policy options;
- reconnecting water to space by providing a spatial planning framework to assist the risks and options of social and economic policy makers;
- to begin to reconnect people to water or their catchment, through the consultative processes of CMPs and the identified partnership options.

In a volume sponsored by the British Hydrological Society it is fitting to return, finally, to the role of science in planning and management. Dovers and Handmer (1995) send us a chilling message: 'Resolution of sustainability issues will remain firmly moral and political, as much as politicians might desire science to produce clear answers, and as much as scientists think they can or should'. It is, however, vital that a balance be set between the new politics of consultation and partnership and guidance which can only come from science, particularly monitoring (during a period of change). What, for example, will be the eventual impacts of the changing relative land-use effects of grass and forest identified by Hudson and Gilman (1993) or the increased dilution provided by raised discharges in the Clyde (Curran and Robertson, 1991)? The widespread 'chain of causation' identified for climate change by Arnell et al. (1994) suggests an urgent need for modelling scenarios at a catchment scale, a task recently taken on by O'Callaghan (1995) but criticised in relation to its direct applicability to the UK planning system by Haslam and Newson (1995). Perhaps it is fitting to end with Dover's and Handmer's term for the role of science in society, in an age of considerable uncertainty: 'ignorance auditing'.

References

Allan, J.A. (1996) The political economy of water: reasons for optimism but long-term caution, in Allan, J.A. (ed.), *Water, Peace and the Middle East*, Tauris, London. pp. 75–117.

Arnell, N. (1996) *Global Warming, River Flows and Water Resources*. Wiley, Chichester.

Arnell, N.W., Jenkins, A. and George, D.G. (1994) *The Implications of Climate Change for the National Rivers Authority*. National Rivers Authority, Bristol.

Bailey, R. (1996) *Water and Environmental Management in Developing Countries*. Chartered Institution of Water and Environmental Management, London.

Brookes, A. and Shields, F.F. (1996) *River Channel Restoration. Guiding Principles for Sustainable Projects*. Wiley, Chichester.

CAG/LUC (1997) *Environmental Capital: a New Approach*. Research & Development Project E2-009, Report E36 (to Countryside Commission, English Heritage, English Nature, Environment Agency). Countryside Commission, Cheltenham.

Curran, J.C. and Robertson, M. (1991) Water quality implications of an observed trend in rainfall and runoff. *J. Inst. Water Environ. Manag.*, 5, 419–24.

Dovers, S.R. and Handmer, J.W. (1995) Ignorance, the precautionary principle and sustainability, *Ambio*, 24(2), 92–7.

EEA/UNEP (1997) Water Stress in Europe: Can the Challenge be met? European Environment Agency, Brussels.

Ellis, J.B. and Gardiner, J.L. (1998) Source control of urban space runoff: a potential for sustainable water resources or recipe for long-term pollution?, in *Proceedings of the ICE/CIWEM Conference, on Maintaining The Flow*. Institution of Civil Engineers, London.

Environment Agency (1997) Hermanus Declaration. *Demand Manag. Bull.*, 25, 2.

Environment Agency (1998) *Thames Environment 21*. Environment Agency, Reading.

European Commission (1998) Amended proposal for a Council Directive establishing a framework for Community action in the field of water policy. COM (97), 49. European Commission, Brussels.

Evans, D. (1997) Assessing the flow reads of rivers. *J. Ch. Inst. Water Environ. Manag.*, 11, 323–28.

Falkenmark, M. (1997) Society's interaction with the water cycle: a conceptual framework for a more holistic approach. *Hydrol. Sci. J.*, 42(4), 451–66.

Falkenmark, M. and Lunqvist, J. (1995) Looming water crisis: new approaches are inevitable', in Ohlsson, L. (ed.), *Hydropolitics: Conflicts over Water as a Development Constraint*. Zed Books, London.

Feldman, D.L. (1991) *Water Resources Management. In Search of an Environmental Ethic*, Johns Hopkins University Press, Baltimore.

Flader, S.L. and Callicott, J.B. (1991) *The River of the Mother of God and Other Essays by Aldo Leopold*. University of Wisconsin Press, Madison.

Gardiner, J.L. (1994) Pressure on wetlands, in Falconer, R.A. and Goodwin, P. (eds), *Wetland Management*. Thomas Telford, London. pp. 47–74.

Gardiner, J.L. (1996) The use of EIA in delivering sustainable development, *Eur. Water Pollut. Control*, 6(1), 50–60.

Gardiner, J.L. (1997a) *Integrated catchment management: experience in the UK and Europe*, in *Proceedings of the 2nd International Conference on Integrated Catchment Management*, Canberra, Australia.

Gardiner, J.L. (1997b) River corridor management and integrated catchment planning for a sustainable water environment, in I. Moreira, G. Saraiva and F. Nunes Correia (eds), *Conservation and Environment Management for Fluvial Systems*. Lisbon.

Gardiner, J.L. (1998) *Thames 21 and Beyond: Towards Sustainable Management for our*

Water Environment Keynote Paper to ASCE Wetlands Engineering and River Restoration Conference, Colorado, USA.

Gardiner, J.L. and Cole, L. (1992) Catchment planning: the way forward for river protection, in Boon, P.J. Calow, P. and Petts, G.E. (eds), *River Conservation and Management*. Wiley, Chichester. pp. 397–407.

Gardiner, J.L. and Perala-Gardiner, C. (1997) Integrating vegetative buffer zones within the catchment, in Haycock, N. (ed.), *Process and Potential of Buffer Zones in Water Protection*. Ardeola Press, Oxford.

Gleick, P.H. (ed.) (1993) *Water in Crisis: a Guide to the World's Fresh Water Resources.* Oxford University Press, Oxford.

Haslam, M. and Newson, M.D. (1995) The potential role for NELUP in strategic land-use planning. *J. Environ. Planning Manag.* **38**(1), 137–41.

Hooper, B., (1997) Breaking down the barriers to effective regional integrated catchment resource management, in *Proceedings of the 2nd International Conference on Integrated Catchment Management*. Canberra, Australia.

House of Commons Environment Committee (1996) *Water Conservation and Supply*, HMSO, London.

Hudson, J.A. and Gilman, K. (1993) Long-term variability in the water balances of the Plynlimon catchments. *J. Hydrol.* **143**, 355–80.

Johnson, S.P. (1993) *The Earth Summit: The United Nations Conference on Environment and Development (UNCED)*. Graham and Trotman, London.

Leeds-Harrison, P.B. and Rounsevell, M.D.A. (1993) The impact of dry years on crop water requirements in Eastern England. *J. Inst. Water Environ. Manag.*, 7, 497–505.

Loucks, D.P. (1997) Quantifying trends in system sustainability, *Hydrol. Sci. J.*, 42(4), 513–30.

Marchand, M. and Toornstra, F.H. (1986) *Ecological Guidelines for River Basin Development*. Report No. 28. Centrum voor Milieukunde, Rijksuniversiteit, Leiden.

Merrett, S. (1997) *Introduction to the Economics of Water Resources. An International Perspective*. University College Press, London.

National Rivers Authority (1991) *NRA Corporate Plan 1992–93*. National Rivers Authority, Bristol.

National Rivers Authority (1994) *Water – Nature's Precious Resource – National Water Resource Strategy*. National Rivers Authority, Bristol.

National Rivers Authority (Thames Region) (1995) *Thames 21 – a Planning Perspective for the Thames Region*. National Rivers Authority, Reading.

Newbold, C., Honnor, J. and Buckley, K. (1989) *Nature Conservation and the Management of Drainage Channels*. Nature Conservancy Council, Peterborough.

Newson, M.D. (1991) Space, time and pollution control: geographical principles in UK public policy. *Area*, 23(1), 5–10.

Newson, M.D. (1992) *Land, Water and Development. Sustainable Management of River Basin Systems*. Routledge, London.

Newson, M.D. (1995) Catchment-scale solute modelling in a management context, in Trudgill, S. (ed.) *Solute Modelling in Catchment Ecosystems*. Wiley, Chichester. pp. 445–60.

Newson, M.D. (1997) *Land, Water and Development. Sustainable Management of River Basin Systems*, 2nd edn. Routledge, London.

Newson, M.D., Slater, S.J. and Marvin, S.J. (1996) *Pooling our Resources. A Guide to*

Catchment Management Planning. Council for the Protection of Rural England, London.

O'Callaghan, J.R. (1995) NELUP – an introduction. *J. Environ. Planning Manag.*, 38(10), 5–20.

Pearce, D. (ed.) (1993) *Blueprint 3: Measuring, Sustainable Development*. Earthscan, London.

Penning-Rowsell, E., Winchester, P. and Gardiner, J.L. (1998) New Approaches to Sustainable Management for Venice. *Geog. J*. In press.

Petts, G.E., (1996) Water allocation to protect river ecosystems. *Regulated Rivers: Res. Manag.*, 12, 353–65.

Postel, S. (1992) *The Last Oasis: Facing Water Scarcity*. Earthscan, London.

Raven, P.J., Fox, P., Everard, M., Holmes, N.T.H. and Dawson, F.H. (1996) River habitat survey: a new system for clarifying rivers according to their habitat quality, in Boon, P.J. and Howell, D.L. (eds), *Freshwater Quality – Defining the Indefinable*. HMSO, Edinburgh. pp. 215–34.

Savory, A. (1991) Holistic resource management: a conceptual framework for ecologically sound economic modelling, *Ecol. Econ.*, 3, 181–91.

Sear, D.A. (1994) River restoration and geomorphology, *Aquatic Conserv.: Mar. Freshwater Ecosyst*, 4, 169–77.

Shuman, J.R. (1995) Environmental considerations for assessing dam removal alternatives for river restoration, *Regulated Rivers: Res. Manag.*, 11, 249–61.

Slater, S.J. (1998) Unpublished PhD Thesis, University of Newcastle upon Tyne.

Slater, S.J., Newson, M.D. and Marvin, S.J. (1995) Land use planning and the water sector: a review of development plans and catchment management plans. *Town Planning Rev.*, 65(4), 375–97.

Smith, C.T. (1969) The drainage basin as an historical basis for human activity, in R.J. Chorley (ed.), *Water, Earth and Man*. Methuen, London, pp. 101–10.

Stansfield, C.B. (1997) The use of water for agricultural irrigation, *J. Chartered Inst. Water Environ. Manag.*, 11, 381–4.

Taylor, A. (1992) *Choosing our Future. Practical Politics of the Environment*. Routledge, London.

UK Round Table on Sustainable Development (1997) *Freshwater*. Department of the Environment, London.

UN Economic Commission for Europe (1993) *Protecting Water Resources and Catchment Ecosystems*. Water Series 1, ECE/ENVWA/31, Geneva.

Walker, J. and Renter, D.J. (1996) *Indicators of Catchment Health. A Technical Perspective*. CSIRO Publishing, Collingwood, Victoria, Australia.

von Weizsäcker, E. (1997) *Factor Four*. Earthscan, London.

Wilby, R.L. (ed.) (1996) *Contemporary Hydrology*. Wiley, Chichester.

Winpenny, J. (1994) *Managing Water as an Economic Resource*. Routledge, London.

Young, G.J., Dooge, C.I. and Rodda, J.C. (1994) *Global Water Resource Issues*. Cambridge University Press, Cambridge.

12

ROLE OF THE BRITISH
HYDROLOGICAL SOCIETY

Frank M. Law

This chapter describes the historical development of the British Hydrological Society (BHS) and its role in providing a forum for debate of scientific and applied aspects of UK hydrology.

The Society is able to cohere the interests of more than 600 professional hydrologists with their three main work sectors across the water cycle of:

- consultancy;
- regulation and management;
- research and education.

Hydrology only gained credibility in Britain as a separate profession in the mid-1960s. The Institution of Water Engineers issued a report (Institution of Water Engineers, 1963) which defined the appropriate post-graduate course qualifications and universities responded with Newcastle and Birmingham following the lead set by Professor Peter Wolf at Imperial College, London. Additional interest came from the International Hydrological Decade initiated (from January 1965) by 57 nations meeting at UNESCO in 1964. It was not long before the Natural Environment Research Council (NERC), and later The Royal Society had a Hydrology Committee. The successor meeting point for governmental hydrology matters is the Inter-Departmental Committee for Hydrology for which the NERC Institute of Hydrology, a component part of the Centre for Ecology and Hydrology, provides the secretariat and chairperson. That committee also doubles as the UK International Hydrological Programme Committee, where hydrologists as a profession are represented by a British Hydrological Society nominee and by the chairperson of the UK Committee of the International Association of Hydrological Sciences.

The BHS was formed in 1982, emerging out of an existing Hydrological Discussion Group based at the Institution of Civil Engineers. From the outset, the Society has been careful to balance the differing emphases of scientists and engineers, hence the special roles granted to the NERC Institute of Hydrology and the Institution of Civil Engineers in the BHS constitution. Its programme themes and publications reflect the changes in hydrological concerns that have emerged over the period of its brief existence (Box 12.1). The society has recognised the need to relate effectively to sister learned societies as water management objectives have become more holistic.

The very diversity of Britain's geology, climate, population density, industry and agriculture means that few global change issues are not exemplified somewhere within its confines. Consultancies based in Britain have long served the wider world, requiring the Society to cover additional trends far beyond a temperate home regime. The Society has a strong regional structure with separate meetings for Scotland, the Pennines, Wales, the Midlands, and southwest England, besides a Southern Section in London that covers East Anglia as well.

The BHS is not yet wealthy enough to give its own research grants, but it did sponsor a broadly-based review of the need for a national hydrological research strategy. Travel grants are given to enable young hydrologists to learn of advances elsewhere, or to take their own findings further afield to a wider audience. The annual post-graduate symposium helps the honing of presentational skills as well as testing the early findings of doctoral research.

The UK Committee for the International Hydrological Programme – a UNESCO activity – asked the Society to organise a key international conference in Exeter University in July 1998. Under the title 'Hydrology in a Changing Environment', the proceedings appeared in a three-volume set (Wheater and Kirby, 1998). The chosen themes reflected the fifth phase of the International Hydrological Programme as well as new issues within the World Meteorological Organisation Hydrology and Water Resources Programme. Because it came at an appropriate point in the science planning cycle, the attendees were encouraged to develop a statement of future needs. A strong consensus emerged and the Exeter Statement has already proved influential; at the Geneva February 1999 Hydrology Conference of WMO/UNESCO an initiative called HELP (Hydrology for Environment, Life and Policy) was agreed.

The Society launched a set of web pages (http://www.salford.ac.uk/civils/BHS/) as early as 1995 and has continued to develop it as an open medium of communication. Its headings reveal its value and the Society's emphasis:

- BHS98, International Symposium on Hydrology in a Changing Environment;
- BHS97, 6th National Hydrology Symposium;
- programme of BHS meetings;
- programme of other hydrological meetings;
- what the Society provides;
- mission statement;

Box 12.1 **Publications of the British Hydrological Society**

Symposium Proceedings

First National Hydrology Symposium, Hull University, September 1987

Second National Hydrology Symposium, Sheffield University, 4–6 September 1989

Third National Hydrology Symposium, Southampton University, 16–18 September 1991, ISBN 0948540303

Fourth National Hydrology Symposium, Cardiff University, 13–16 September 1993, ISBN 0948540524

Fifth National Hydrology Symposium, Heriot-Watt University, Edinburgh, 4–7 September 1995, 412 pp. ISBN 0948540729

Sixth National Hydrology Symposium, University of Salford, Manchester, 15–18 September 1997. ISBN 0948540826

Occasional Papers

BHS Occasional Paper No. 1 (1988) An introduction to operational control rules using the 10-component method, A. Lambert, 45 pp

BHS Occasional Paper No. 2 (1989) Weather radar and the water industry, 91 pp. ISBN 0948540184

BHS Occasional Paper No. 3 (1994) Assessment of drought severity. Ed. J. Mawdsley, G. Petts and S. Walker. 30 pp. ISBN 0948540605

BHS Occasional Paper No. 4 (1994) Analytical techniques for the development and operations planning of water resources and supply systems. Ed. T. Wyatt. 100 pp. ISBN 0948540710

BHS Occasional Paper No. 5 (1995) Hydrological uses of weather radar. Ed. K.A. Tilford. 164 pp. ISBN 0948540680

BHS Occasional Paper No. 6 (1995) Modelling river-aquifer interactions. Ed. P.L. Younger. 165 pp. ISBN 0948540702

BHS Occasional Paper No. 7 (1996) Palaeohydrology: context, components and application. Ed. J. Branson. 75 pp. ISBN 094854077X

BHS Occasional Paper No. 8 (1997) Floodplain rivers: hydrological processes and ecological significance. Ed. A.R.G. Large. 130 pp. ISBN 0948540818

BHS Occasional Paper No. 9 (1998) Hydrology in a Changing Environment: BHS international conference poster papers and index, Exeter University. ISBN 0948540877

Miscellaneous

The BHS Research Strategy: Sustainability in a changing world: the key role of hydrology. 18 pp. ISBN 0948540605.

The Next 10 Years. 4 page leaflet. Free. (Available as a downloadable Postscript document from the BHS-hydrology electronic mailing list.)

Box 12.2 **The Exeter Statement, 10 July 1998:**
International conference on Hydrology in a
Changing Environment, University of Exeter

CONTEXT

Experts and the public are both more troubled than ever before about both the future for water management and the protection of environmental systems. In response to that situation, the International Conference on Hydrology in a Changing Environment drew 250 attendees from seventy nations to consider 164 papers on key themes of the UNESCO International Hydrology Programme and the WMO Hydrology and Water Resources Programme.

The six conference themes over the week from 6–10 July were a reminder of those concerns:

• global hydrological processes;
• ecological and hydrological interactions;
• groundwater at risk;
• hydrology of environmental hazards;
• catchment management and resource assessment in dry areas;
• hydrology of large cities.

The Conference recognised that hydrologists are being asked to work on a wider canvas. Integrated or holistic management of river basins is involving ever larger contacts with other stakeholders to examine the environmental significance of the various hydrological pathways, rivers being the most obvious.

Interaction with economists, lawyers, conservation and community groups, consumer interests, industry managers, navigation bodies and many more is known to be necessary. The dilemma is whether the legitimate need to improve hydrological science, and to disseminate its results, will get lost in a 'democracy' of other concerns. The need to improve hydrological understanding in depth, and to reduce the uncertainty of its predictions, could become marginalised whereas it should be the cornerstone of effective basin management.

Research success in the field of global hydrological processes (ongoing since the 1982 IAHS Assembly at Exeter) has contributed to motivating international agreements on climate change. It is known that:

• water is the primary greenhouse gas;
• the fate of increased water in the atmosphere is the primary cause of uncertainty in climate change predictions;

- input by hydrologists of observations, process understanding and sub-models has contributed to improving credibility of General Circulation Models;
- calibrated coupled models of surface energy and water balance for vegetation prone to change has facilitated predictions of likely alterations in regional and global climate.

Hydro-ecology has come of age in the last decade, with physical habitat simulation modelling in wide use for policy setting. The interplay of biology with hydrology is a major theme for the new millennium.

Numerous case studies from all continents had demonstrated to attendees that success could be achieved in specific projects to master long-running environmental or resource scarcity problems.

The organisers of the Exeter Conference sought primarily to give those present (many of them senior figures in hydrology in their home countries) an opportunity to identify leading issues and critical opportunities which should:

- drive national and continental research in the new millennium
- assist the Fifth Joint UNESCO/WMO International Conference on Hydrology, set for 8–12 February 1999 in Geneva, to make useful progress in co-operative activity.

KEY IDEAS FROM THE THEMED DISCUSSIONS
These often had much in common and hence are given jointly here:

- Better catchment management is often hindered in much of the world by hydrometric records that are too short to capture the variability against which society now seeks protection and timely warning. This is particularly obvious in arid areas but our changing environment makes the urgency for sustaining long index records far more general than that.
- We need to communicate our extensive experience, while continuing to listen to those needing it. Dialogue must be:

    between people within our discipline and outside it;
    between specialists and water managers;
    between specialists in and newcomers to hydrology;
    between specialists and the public.

Meanwhile the transfer of educational knowledge must be sustained.
- We recommend that consideration be given to a Second International Hydrological Decade as a major project within the current framework of activities of UNESCO, WMO and others. It would recognise the existing world observation programmes in related sciences, and be targeted at providing those comprehensive datasets and interpretative science needed to lower the uncertainty in hydrological prediction in

274

areas of environmental, economic and social importance. It could build upon time series, for the 2001–2011 period, of variables as diverse as:

> global surface moisture;
> continental soil–vegetation–atmospheric transfers;
> digital water tables of regional aquifers;
> city water input, loss and effluent time series.

- Over-used resources threaten many people and often only partial solutions have been found; alternatives to conventional water supply and waste disposal are still needed for some countries.
- Society will benefit by hydrologists demonstrating effective operational rules based on models that incorporate predictive uncertainty in all elements.
- Professionals must strike a balance of effort between modelling and field data collection, as poor models threaten professional credibility.
- Forecasts are only as useful as the ability of the audience to interpret them.
- Progress will be helped by freer data exchange; but it is recognised that some regions have no records and others have them for short time spans that could be unrepresentative.

IDEAS MORE SPECIFIC TO EACH THEME
Those ideas held with conviction are listed below.

Global hydrological processes
- There is a need for a better understanding of long-term and global scale patterns and processes to aid the production of seasonal and regional scale forecasts.
- Land/atmosphere/ocean coupled models need to be developed further, and extended to include to water quality and ecology problems.
- More physically based methods for downscaling global predictions to local conditions need to be developed.

Ecological and hydrological interactions
- The need is for multi-disciplinary teams, unfettered by the limits of the original education of the different scientists; hydrologists venturing into ecology in isolation will have difficulties.
- Research should range over scales including catchment, river reach and microhabitat levels, in environmental zones from headwaters to river mouths.
- Similar research projects will be required in different areas where biotic species assemblages exhibit clear differences.
- Biodiversity protection requires our contribution, with advancement from simpler interpretations to the more complex unsolved mysteries.
- River bed flux research is a promising new field.

Groundwater at risk
- The 4-dimensional environment demands fundamental process understanding, directed now towards recharge and unsaturated zone processes, and aquifer/stream water interactions.
- Recharge estimation is held back for lack of a satisfactory way of measuring bulk vertical conductivity.
- Planners and the public should be urged to support the control of aquifer use, implemented by strong institutions.

Hydrology of environmental hazards
- Our ultimate concern is to reduce the likelihood of disasters, concentrating on reducing vulnerability both by works, land zoning and early warning systems.
- Weather hazards feed through to ecological systems in ways not well understood.
- Greater integration of hydrological and climatological aspects of hazards should be featured, along with uncertainty.
- Constraints to use of hydrological warning systems are often social or economic.
- Models of landslides, as an example phenomenon, did not often have calibration data.

Catchment management and resource assessment in dry areas
- Quantifying the high variability in time and space of hydrological variables, including sediment loads and flood transmission losses.
- Understanding the interactions between:

  ephemeral flows and alluvial groundwater;
  shallow and deeper groundwaters;
  land use, sediment generation and surface storage management;
  scale dependence of runoff and channel transmission losses.

- The profession needs to draft, in conjunction with other competent institutional bodies, enforcement measures to control and protect water resources in arid areas; this will need to be done locally, using best practice from other related areas.

Hydrology of large cities
- Urgent needs exist for integrated systems modelling and management, focused on interactions of components.
- Regulations related to city hydrology have been promulgated (with good intent) ahead of the scientific and technical capability to deliver a quantified solution in many cases e.g. water quality monitoring.
- Both active (hands-on) high technology and passive (non-structural) approaches are desirable; post-implementation maintenance cannot be neglected.

- Transplanting either of the above approaches to cities in developing nations must be questioned.
- Radical rethinking is warranted for wastewater separation at source, with recycling of greywater for other uses.

CONCLUSION
After revision by plenary discussion, the conference attendees adopted this Statement by acclamation at their final session, and asked that the British Hydrological Society put it forward through the UK Government to the organisers of the 1999 Inter-Governmental Conference on Hydrology, Geneva.

- organisation and regional information;
- societal associations;
- who's who at BHS;
- honourary members and past presidents;
- contact addresses;
- publications;
- education and training;
- electronic mailing list (discussion group);
- awards and prizes;
- membership analysis;
- register of hydrological consultants;
- links to related WWW sites;
- hydrological situations vacant;
- how to join;
- Chronology of British Hydrological Events.

It has fostered an unusual, and growing, set of WWW pages called *The Chronology of British Hydrological Events*. This recognises the need for public understanding of environmental variability, with or without climate change fluctuations. Hence it gives a chronology of hydrological facts across Britain. Search tools enable those with internet access to scan for a particular phenomena, or location, or river, and the source reference. It is hoped that such public domain contributions will encourage other disciplines with an interest in environmental history to follow suit.

Response to change is sometimes quasi-political. The BHS submitted evidence to the House of Lords Select Committee on Sustainable Development (House of Lords, 1994) and its concern for over-development of aquifers received due attention in their Lordship's text. At other times pressure can be applied 'behind the scenes' in government circles, albeit limited by the need for consensus within the activist core of the Society. A current example of topics BHS members have raised for attention by legislation is the need for better rules to govern the provision of balancing storage in urban developments, and the desirability of ensuring that all headwater streams come under the maintenance programme of the Environment Agency.

English resorts have long had an index of their summer weather (Poulter, 1962) but achieving water industry consensus about (or funding for) the equivalent of the Palmer drought index used in the USA (Palmer, 1965) has been elusive. *BHS Occasional Paper 3* was a compromise struck in 1994 but its attempt at the equivalent of a Richter scale for drought has failed to make an impression. This is attributable at one level to a lack of thorough scientific research and at another level to the intangible variations in drought as viewed by different water users. Instead, the 1975–76, 1984 and 1988–92 droughts have each received individual attention (Central Water Planning Unit, 1976; Marsh and Lees, 1985; Marsh *et al.*, 1994) and strenuous efforts have been made by the Environment Agency to improve the hardest hit of 40 low-flow rivers (National Rivers Authority, 1993).

Floods are more dramatic and it has been a tradition, possibly dating largely from the 1771 flood that was so outstanding in Northern England (Archer, 1992), to install a floodmark to record events unlikely to be surpassed for at least a generation. Examples can be found in masonry, brick, cast iron, wood and plastic. Some form the core of chronologies (McEwen, 1990, 1987) while others get a passing mention or are buried deep in reference literature (Griffiths, 1983). Regrettably the Royal Commission for Historic Monuments takes no formal interest in them and there is no book that covers their existence. Even Thames Path literature fails to highlight the metal plates marking the November 1894 peak (roughly a 100-year flood) or the painted wood marks of March 1947 (about a 50-year flood) at lock-keepers' buildings. The BHS has been trying to find funding for a WWW catalogue of these pointers to the risks that are run by at least 6% of the nation's population (Morris and Flavin, 1996) who live either in partially defended or undefended floodplains.

The Society works at the interface between the earth sciences and life sciences. It is never solely engineering or technology or science. There are members across the spectrum from pure to applied science. All are held by their involvement in the water cycle, initially in terms of resource quantities, then addressing chemical quality and most recently looking at the life forms that are interwoven with water pathways. The BHS has chosen quite deliberately not to be exclusive in its membership, nor has it sought status for its members. However, it does have aspirations to be taken ever more seriously in planning and political circles. To that end, its meeting topics are chosen for their relevance to live or coming issues, and its Committee will ask a representative to respond to the more important government consultation documents where professional hydrologists have a known view.

The Society has a training statement and exists both to share experience and to encourage the highest standards among its member hydrologists. Those with an appropriate degree can become chartered civil engineers but a larger number become corporate members of the Chartered Institution of Water and Environmental Management. However, some specialist hydrologists may wish to gain their professional qualification through the Royal Meteorological Society or the Geological Society. Continuing professional development is the community's guarantee that the changing hydrology of our global village will not take any thoughtful person by surprise.

References

Archer, D. (1992) *Land of Singing Waters: Rivers and Great Floods of Northumbria.* Spredden Press, Stocksfield.

Central Water Planning Unit (1976) *The 1975–76 Drought: a Hydrological Review.* Technical Note 17. CWPU, Reading.

Griffiths, P.P. (1983) *A Chronology of Thames floods*, 2nd edn. Water Resource Report. Thames Water Engineering Directorate.

House of Lords (1994) *Minutes of Evidence taken before the Select Committee on Sustainable Development.* Questions 1196 and 1276, 22/11/1994.

Institution of Water Engineers (1963) Report of the committee on the education and training of hydrologists. *J. Inst. Water Eng.*, **XVII**, 381–91.

Marsh, T.J. and Lees, M. (1985) *The 1984 Drought*. Report in Hydrological Data UK Series. Institute of Hydrology/British Geological Survey, Wallingford.

Marsh, T.J. *et al.* (1994) *The 1988–92 Drought*. Report in Hydrological Data UK Series. Institute of Hydrology/British Geological Survey, Wallingford.

McEwen, L.J. (1987) The use of long-term rainfall records for augmenting historic field series: a case study on the upper Dee, Aberdeenshire, Scotland. *Trans. R. Soc. Edinburgh*, **78**, 275–87.

McEwen, L.J. (1990) The establishment of a historical flood chronology for River Tweed catchment, Berwickshire, Scotland. Scottish Geog. Mag., **106**, 37.

Morris, D.G. and Flavin, R.W. (1996) Flood risk map for England and Wales. *Institute of Hydrology Report 130*. Institute of Hydrology, Wallingford.

National Rivers Authority (1993) *Low Flows and Water Resources: Facts on the Top 40 Low Flow Rivers in England and Wales*. National Rivers Authority, Bristol.

Palmer, W.C. (1965) *Meteorological Drought*. Research Paper 45, US Weather Bureau, Washington, DC.

Poulter, R.M. (1962) The next few summers in London. *Weather*, **XVII**, 253–5.

Wheater, H. and Kirby, C. (eds) (1998a) *Hydrology in a Changing Environment, Vol. I.* Proceedings of the British Hydrological Society International Conference, Exeter, July 1998. Wiley, Chichester.

Wheater, H. and Kirby, C. (eds) (1998b) *Hydrology in a Changing Environment, Vol. II.* Proceedings of the British Hydrological Society International Conference, Exeter, July 1998. Wiley, Chichester.

Wheater, H. and Kirby, C. (eds) (1998c) *Hydrology in a Changing Environment, Vol. III.* Proceedings of the British Hydrological Society International Conference, Exeter, July 1998, Wiley, Chichester.

13

FUTURE UK HYDROLOGICAL RESEARCH

Jim Wallace, Enda O'Connell and Paul Whitehead

This chapter identifies the key gaps in hydrological knowledge which will need to be filled in order to address the major water-related issues facing the UK today and in the foreseeable future.

13.1 Introduction

The focus of hydrological research in the UK tends to change with the trends in national and international water-related issues. When there is either too much or too little water, research tends towards aspects of water quantity. Whenever water appears to be available in adequate quantities or when its chemical properties pose a threat to human or environmental health, hydrological research tends towards water-quality issues. For example, 30 years ago national concerns about water supply led to hydrological research into the effects of afforestation on the water yield to our upland reservoirs. During the 1980s, international concerns about acidification of rivers and lakes increased the emphasis on water quality research. The period of exceptionally dry and wet years in the late 1990s in the UK has brought the problems of water quantity back to the fore and clearly demonstrated that even in the UK fresh water is neither an inexhaustible resource nor an insignificant threat to life and property. The challenge to UK hydrologists is to ensure we maintain a robust and well-balanced programme of research which will provide the scientific basis to address current and future water-related issues. This means we need well-designed basic research into four principal areas of hydrology: water resources, extreme events, water quality and the bio-physical processes which control the movement of water within the hydrological cycle.

Water resource issues are basically a question of supply and demand. How much water is there in a given place at a given time and how does this compare with fresh water demands? Even in the UK there are clearly regions where demand can sometimes outstrip supply. Assessment of water supplies in our reservoirs, rivers and aquifers is dependent on hydrological models, many of which still need to be improved to provide results of acceptable accuracy. For example, the rate of recharge to our aquifers, especially those in the south of England that supply much of the local water demand, is not well known. Current demands are largely domestic, agricultural and industrial. However, a major change to the supply/demand debate has arisen due to the modern expectation that fresh water will be available in sufficient quantities, not only to meet human needs directly, but also the full range of environmental needs. The latter include water requirements to support ecological systems, such as the plants and animals in wetlands and fish and invertebrates in rivers and lakes. Exactly how much water do wetland ecosystems require? What are the consequences of sub-optimal supply?

Extreme hydrological events, such as floods and droughts, can still produce devastating effects in the UK. We had barely stopped discussing the publicly unacceptable water shortages of the 1995 and 1996 summers, with their widespread effects on water supply, river flows and aquatic ecology, when the exceptional rainfall across the midlands at Easter 1998 led to the worst flooding and damage that those regions had ever recorded. Exactly how exceptional where these events? Has the frequency with which they may re-occur changed and if so why and to what?

Water-quality research is driven by the growing concern about the degree to which we may be polluting our fresh water resources, thereby jeopardising their ability to sustain human and ecological requirements. National and European drinking and river water standards are generating a need for more information on the chemical composition of these waters and the sources and pathways by which pollutants enter them. Much aquatic pollution is transported by sediments, but we do not yet know enough about this important process. Many pollutants occur at very low concentrations and remain undetected and/or unidentified. There is therefore a need to develop and apply methods to detect and identify these chemicals as well as their toxicity to human and ecological health.

Water plays a key role in many biological and physical aspects of the 'Earth system'. For example, water is an important factor in climate change and variability. With the current focus in the global warming debate on carbon dioxide, it is often forgotten that water vapour in the atmosphere is the single most important radiatively active gas. In the current estimates of the warming effects of higher atmospheric carbon dioxide concentrations, about half of the warming comes from the consequent higher atmospheric water vapour concentration. Atmospheric water vapour is also very dependent on the land surface conditions which control the rate of evaporation. Greater water vapour concentrations in the atmosphere may also lead to increased cloudiness, which cools the Earth's surface and atmosphere. Therefore, the net effect of an increase in atmospheric carbon dioxide concentration cannot be predicted without an adequate description of the hydrological processes affecting climate.

The following section elaborates on some of the key research questions that the above national and international issues raise. These are presented under the four principal hydrological research areas identified above, i.e.

1. *bio-physical processes* that control the mass transfer of water within the hydrological cycle;
2. physical, chemical and biological processes that affect *water quality*;
3. *water resource* assessments which quantify the availability of fresh water at space and time scales suitable for effective water management; and
4. the estimation of *hydrological extremes* for human and environmental protection.

13.2   Bio-physical processes

Bio-physical processes need to be studied across a range of scales. At the smallest scale (millimetres to metres), the importance of sub-surface processes is becoming increasingly recognised. These control the net exchange of energy and water with the atmosphere, rivers and aquifers. Specific new areas for research include the study of how soil heterogeneity (variability, cracks, macropores, etc.) and soil–root interactions affect key components of the water balance (i.e. evaporation, runoff and drainage). At a field scale, since most of the world's vegetation contains mixtures of species, we need to focus more attention on the partitioning of water in the complex mixed plant communities that occur in nature, e.g. in heathlands, wetlands and riparian zones. At a complete landscape scale, the challenge is to deal with the typical patchwork of different land uses. We need to understand more about how water flows between different landscape units, such as forests, heaths, crops, set aside and pasture, if we are to be able to predict the consequences of changing the composition of the landscape. For example, headwater grasslands and forests may help to reduce runoff during wet periods, increase filtration to the soil and aquifers and reduce soil erosion.

A major emerging international issue is concerned with the water required to grow food for the increasing world population. Current projections estimate that by 2025 most of the world's population may not have sufficient water to grow their basic food requirements. This enormous food gap seems unavoidable in water-scarce regions unless more efficient use can be made of existing water resources. A major effort is therefore required to look at the technical and non-technical aspects of increasing the efficiency with which water is used in both rain fed and irrigated agriculture. To ensure that any consequent impacts on downstream (and/or upstream) water users is taken into account, this work needs to be carried out within a catchment framework.

At the largest continental to global scale, the UK hydrological community has been a key player in international global change research, through both steering and executing the science. In collaboration with our meteorological colleagues, we have invested significant resources to develop this national capability and we need to capitalise on these strengths by continuing to contribute to research into how large-scale land–atmosphere interactions affect climate. The most pressing need is for the development of multi-source and combined energy, water, carbon dioxide flux and plant growth models which can operate within the constraints of Global Circulation Models (GCMs). There is also a need to improve the representation of runoff within GCMs to meet the growing demand for climate impact assessments. The data which are essential for the testing and development of these models will have to come from a combination of well-instrumented hydrological catchments, well-designed experiments and improved use of remote sensing techniques.

International emphasis in climate research is shifting in two principal ways. Firstly, there is a move from regions subject to tropical deforestation and desertification to northern latitude biomes, such as boreal forests and tundra, where global warming impacts may be greatest. Secondly, the focus is shifting from climate change to climate variability and its human and environmental impacts (e.g. via flooding, drought and agricultural production). Climate variability is strongly linked with changes in sea surface temperatures, for example, in the South Pacific (*el Niño*) or North Atlantic (North Atlantic Oscillation). To tackle the emerging issue of the possibility of mechanistic connections between sea-surface temperature anomalies and weather conditions over the land surface, it will be necessary for much closer collaboration to be forged between hydrologists and oceanographers. Since UK weather is often dominated by North Atlantic sea conditions, this liaison is vital to addressing the future national focus on the study of the impacts of climate variability on land use and water resources.

13.3 Water quality

Specific challenges in water-quality research are driven by concerns about fresh water pollution and legislation on drinking water quality. Underlying this are issues to do with human health and an enhanced expectation for ecosystem conservation. Rising population levels and the need for clean water in developing countries has also focused interest, and hence, research on water quality issues. Trans-boundary air pollution and the issue of acid rain gave a considerable boost to water quality research in the 1980s as it was realised that hydrological process knowledge had to be combined with an understanding of chemical processes in order to explain observed acidification behaviour in soils, lakes and streams. Water-quality research has also been stimulated by groundwater pollution problems and this has led to a better understanding of in-stream pollutant mixing processes. The movement and distribution of micro-pollutants, such as pesticides, in the soil and groundwater system is also of increasing concern. These micro-pollutants occur at very low concentrations and many are unidentified and have unknown consequences for human and ecological health. The health of aquatic eco-systems may also be being affected by oestrogenic compounds in rivers. There is, therefore, a need to understand the impacts of different concentrations of oestrogen on in-stream ecology and to identify critical levels to set water quality objectives. All of the above issues highlight the generic need to under-stand the processes controlling the fate and transport of pollutants, both across the landscape and within our rivers and aquifers.

At the complete catchment scale, water-quality modelling faces a similar challenge to physical process modelling at this scale, i.e. the development of spatially distributed water quality models which can cope with large hetero-geneous areas. Clearly, these models cannot be developed without appro-priate measurements of how water quality varies in time and space. A key challenge is to understand how the larger-scale catchment behaviour is re-lated to the heterogeneity observed at smaller scales. This problem of scaling up localised information was one of the tasks of the LOIS (Land–Ocean Interaction Study) project. Progress was made towards generating integrated catchment models for water quality. However, LOIS has also identified that the complexity of the physical and chemical processes and difficulties of integrating in both space and time means that there is much left to do in this area. For example, much of the pollution in our aquatic environment is transported by sediments and so we need to understand more about the sources and movement of sediments throughout the entire catchment system. It is now also recognised that most pollutants in catchments, rivers and lakes are directly affected by biological processes. For example, nitrogen trans-formations are controlled by micro-biological processes and it is essential to incorporate these in water-quality models in order to predict the short-term dynamics of nitrogen in aquatic environments. Organic compounds (e.g. pesticides) directly affect and are affected by biological systems as are other nutrients, such as phosphorus or metals, in the natural environment.

The importance of the interactions between water quality and aquatic biology is highlighted by the issue of how much oxygen is dissolved in water. Dissolved oxygen in rivers and lakes is essential for supporting healthy aquatic ecosystems. The amount of oxygen in solution is controlled by a combination of factors: physical factors, such as turbulence processes and re-aeration; chemical factors, such as pollutant transformation; and biological factors, such as phytoplankton and macrophyte growth. The dynamics are very complex and interactions between physical, chemical and biological systems will, therefore, be a major area of research for the future. The short-term dynamics of dissolved oxygen linked to phytoplankton and macrophyte behaviour is of particular concern.

Another important area of water-quality research is associated with the use of 'natural' filtration systems to remove pollutants from industrial or domestic sources. For example, this is a technique being used to deal with acid mine drainage. In these systems, wetlands or passive treatment systems

Plate 13.1 Collecting river flow velocity data for hydro-ecological research (Mike Acreman).

are currently being designed and built to treat waste from abandoned mines. Very little is known about the micro-biological processes controlling the behaviour of these systems and wetland schemes are being built on the basis of relatively thin knowledge. In general, there is increasing interest in what are seen as more natural and environmentally acceptable forms of waste water treatment. We, therefore, need to assess the degree to which wetland ecosystems can improve water quality through removal of nutrients, sediment and pollutants, so that improved water management can be achieved through working with nature rather than against it. This will require the combined skills of hydrologists, chemical engineers and freshwater ecologists to provide the underpinning scientific knowledge needed to design and manage these schemes.

Plate 13.2 Measuring water levels in a dip-well on the North Kent Marshes (Mike Acreman).

There is also an important new role emerging for water quality and water quantity research in resolving local, national and international conflicts. The understanding of complex water quality and quantity issues that quantitative hydrological research tools provide can assist in determining a lasting resolution to conflicts. A particular challenge is the allocation of water resources to meet the often conflicting objectives for our rivers, lakes and wetlands which may be economic (e.g. provision of domestic tap water at a reasonable cost), ecological (e.g. conservation of specific river habitats and species), legal (e.g. maintenance of existing abstraction rights) or aesthetic (e.g. scenic beauty). For example, in the UK, local disputes, such as water abstraction in the River Kennet and the perceived long-term decline of water quantity and quality in this river, have led to a comprehensive set of hydrological studies considering stream flow, water quality and hydro-ecology. These have been central to the resolution of disputes over the effects of water abstraction on river flow and fish populations. This need to provide sound scientific information for conflict resolution will continue to act as a major stimulus to water quality and quantity research at local, regional, national and international levels. However, future research in this area needs to be carried out in collaboration with social and political scientists to define methods to establish clear objectives for setting appropriate river flows and wetland water levels, such as through stakeholder participation and the development of catchment management plans.

13.4 Water resources

In the field of water resources, there is a need for more accurate information on how much water there is in a given area at a given time at spatial and temporal scales that are of use to water resource managers. Sustainability remains a key issue in this regard: how can the sustainable yields of reservoirs and aquifers be defined so as to achieve the delicate balance between supporting economic development and lifestyle, and protecting the water environment from irreversible damage? Progress towards attaining sustainable water resources management policies needs to be underpinned not only by research on the impacts of abstraction on the bio-physical system, but also on the key socio-economic issues associated with demand. In order to develop the most useful water resource management tools, hydrologists therefore need to be proactive in widening their scope to include non-hydrological drivers and impacts. For example, ownership and the real or implicit cost of land and water have a profound effect on how they are used. No sustainable water management solution can be evolved which ignores these factors. Being an inherently multi-disciplinary science, hydrology has the basic structure to take a lead in developing the holistic approach which is needed to solve many future water-management problems. Ultimately, the use of all commodities is controlled by legal regulation and hydrologists should have the confidence not only to provide information to pursue solutions within given legal frameworks, but also to provide the information needed to formulate new water laws.

We also need to concentrate more effort on the development of consistent and reliable ways of estimating groundwater recharge over large areas. There is a particular need for this in the lowland permeable catchments of the south and east of the UK, since this is where water demands are closest to the supply which is highly dependent on groundwater. This subject is particularly important given the recent interest in reassessing the UK's fresh water resources. A key impediment to progress in this area is the lack of understanding of mechanisms controlling surface–groundwater interactions. Hydrologists need to collaborate more closely with hydro-geologists in order to make progress in this important area. On a European and wider international scale, regional and global fresh water resources are very poorly quantified and so we need to put greater effort into the collation of hydrological data and modelling that is required to improve water resource assessment at these scales.

The exceptionally variable climatic regime observed in recent years in the UK and mainland Europe has raised difficult questions for hydrologists and water resources managers alike. Is our climate changing sufficiently to affect water resources and the demands on them? Observational records, and the available statistical tests, cannot generally provide conclusive evidence of climate change, primarily because of the masking effect of 'natural' climatic variability. This variability may in itself be changing. The net result is that hydrological records are not sufficiently long to define analogue periods in the past which might be similar to that observed recently in the UK.

As to how the climate may evolve in the future, GCMs are the only predictive tools available, and much research effort is currently being devoted to the development of methods for 'downscaling' GCM output to give realistic space-time rainfall scenarios at the catchment scale. Recent developments in space-time rainfall modelling have enabled improved methods of downscaling and scenario generation to be developed. However, while GCMs can reproduce reasonably well the main features of the current climate system on a global scale, deficiencies on regional scales are still evident which can introduce significant errors into downscaled scenarios. These need to be taken into account when GCM scenarios for future climates are employed in impact studies.

Even if realistic space-time rainfall scenarios could be generated for impact assessment, these still need to be converted into river flow and aquifer recharge scenarios for use in water resource assessment studies. Moreover, since land use change over aquifer outcrops/within catchments can significantly alter recharge/runoff regimes over time, the models used for converting the rainfall scenarios into recharge/runoff scenarios need to be able to accommodate such changes. Physically-based distributed models have the structure needed to accommodate such changes, but their data and computational demands have restricted their widespread use. However, new developments in this area should enable simplified physically-based modelling frameworks to be applied to larger catchment areas.

While hydrological models can always be calibrated for catchments with available discharge records, applications to ungauged catchments will require new regionalization methods. Developments in remote sensing and Global Information Systems (GIS) technologies can be of major assistance here. However, reliable quantitative links between the model parameters and catchment topography, soils, vegetation and geology will be needed if viable regionalisation methods are to be developed. Physically-based model parameters are desirable in this regard, but applications on large scales will provide a major challenge. Methods are needed of classifying catchments into classes with hydrologically similar responses, sometimes called Hydrological Response

Units (HRUs), and of linking small-scale physical parameter measurements with these HRUs. This will provide the all-important link with small-scale bio-physical process research, and enable predictions to be made at catchment scales which are properly sensitive to changes in land use.

13.5 Hydrological extremes

Floods and drought continue to be the natural hazards which kill and affect most people world-wide. In the UK, floods cause major disruption, damage and associated financial losses. Accurate flood and drought estimation is therefore still a major national and international issue. Improved flood estimation may be achieved by moving on from conventional design-event methods to continuous simulation techniques. Developments in space-time rainfall modelling and in distributed catchment modelling should eventually enable long catchment runoff series to be synthesised. This will have the advantage of improving rare-event estimation and will also allow the potential impacts of future climate change on hydrological extremes to be more readily assessed.

Extreme-event estimation also needs to be taken into account the problem of non-stationarity in hydrological time series (rainfall and river flows). Current methods of estimating flood return periods assume no long-term trend. This is acceptable in a stable climate satisfying the conditions for stationarity, in which case the return period is a valid measure for assessing levels of flood protection. However, under transient, non-stationary conditions, such a measure is no longer valid, as is evident when it becomes increasingly difficult to explain an unusual sequence of climatic events as attributable to chance alone. Measures of risk need to be developed which are appropriate to transient conditions that can quantify any changes in hydrological variability which may be occurring.

The opposite hydrological extreme to floods, i.e. low flows, can also have serious water resource and environmental impacts. Low flows in rivers are occurring more frequently due to the combined effects of low rainfall and groundwater abstraction. This can have important effects on water supply, wetlands and aquatic ecology. To address this issue hydrologists who are experts in estimating low flows and drought will need to work more closely with fresh water ecologists to research this important interface between hydrology and ecology. This is a separate but complementary requirement to the need to understand the impacts of water quality on freshwater ecology, which was identified earlier.

Both extreme event and water resource assessment need to address the issue of uncertainty. No data and/or model is without error and so flow frequency estimation and forecasting and water resource estimation need to have their uncertainties properly identified if they are to be reliably used for flood defence and resource management purposes. These uncertainties also need to be incorporated into the decision-making tools provided to flood defence and water resources managers. More realistic management decisions can be made by acknowledging these uncertainties rather than ignoring them. Hydrologists could also improve the presentation of their results and models so that they are more easy for decision makers to understand and use.

13.6   Concluding remarks

Whatever the future environmental issues involving water may be, some basic skills are required by the hydrological scientists who will be expected to address them. The definition of these skills is best learned from the history of the successful developments in hydrological science. Here most progress has been made by the combination of observation – making measurements of precipitation, evaporation, runoff or drainage – with hypotheses, usually in the form of mathematical descriptions ('models') which may explain the observations. This tells us several important things. Firstly, that instruments continue to be needed to allow the necessary measurements to be made. For example, the neutron soil moisture probe was developed by hydrologists to enable the reliable and routine measurement of the water content of soils. The vast amount of data generated by the use of this instrument world wide has increased our understanding of the water use of a wide range of crops and trees in many environments and has led to the development of practical models of their water requirements and growth.

Future instruments are, therefore, needed to allow measurements of key hydrological variables, generally in the places and at the scales where they are most difficult to measure. Rainfall data are required over large areas and solid precipitation is often inadequately measured. Evaporation accounts for two-thirds of all terrestrial rainfall, but is still difficult to measure directly over heterogeneous terrain, (e.g. in sparse canopies, hilly terrain, complex patchworks of vegetation, soil and water). Soil moisture measurements are required at both small and large scales. At small scales, e.g. in the vicinity of plant roots, soil moisture is required to better understand the sub-surface control of evaporation (i.e. as transpiration). Soil moisture is also required over large areas for large-scale hydrological and climate modelling. Runoff measurements are needed over large heterogeneous areas and, in particular, information is needed on the effects of run-on and runoff on in landscapes containing a patchwork of different soils and land uses. Drainage continues to be particularly difficult to measure directly, but is important for the estimation of recharge to aquifers, base flow in rivers and the transport of pollutants.

Having the instruments alone is not sufficient. It is also necessary to be able to deploy and maintain these devices in the many field environments where data are required. The skill and experience to do this is far from trivial and as the complexity, harshness and remoteness of the environments where data are now required becomes more acute, good field experimental skills need to be even more highly developed. Training of hydrological scientists in practical field research is, therefore, a fundamental and constant requirement of future hydrological science.

Once good reliable observations have been obtained, hypotheses concerning the processes which control the rates of exchange of water and its chemical composition within the hydrological cycle can be tested and models of the system developed. Strictly speaking, a hypothesis or model can never be proved or 'validated', it can only be disproved or invalidated. More progress would be made in hydrology by scientists seeking to understand the observations which their current model *cannot* explain. They can then make the minimum change to their theory or hypothesis to account for the new observations and proceed to obtain further data that the revised model will again no longer explain. The final essential skill required to advance hydrological science is, therefore, the ability to develop hydrological models. This requires a sound understanding of the underlying physics, chemistry and physiology and the mathematical ability to describe the controlling processes in appropriate models.

Observational techniques, the skill to use these properly in the field and the ability to synthesise the results in models of the system under study are the fundamental elements which have sustained the development of hydrological science so far and will continue to be the basic skills which will sustain this important science in the future. Significantly, they are invariant with fashions in environmental issues and, therefore, guarantee both real scientific progress as well as the ability to deal with whatever issue is to be addressed in the future.

INDEX

Note: page numbers in italics refer to figures and tables; those prefixed by a 'B' refer to boxed material.